FREETH, R. Lighting: plan your home.

1970.

7

Plan your home

LIGHTING

Richard Freeth

Studio Vista

I should like to thank Paul Yule without whom this book would never have been begun, let alone finished, for all his help in ways too numerous to mention. Also my thanks are due to A.D.S. Atkinson for his technical assistance.

Photographs by courtesy of the Lighting Industry Federation and its members except where otherwise credited.

Cover photograph
Architect: Jack Bonnington

The opinions expressed in this book are the author's and not necessarily the official views of either the publisher or the Lighting Industry Federation.

Published 1970 by Studio Vista Limited
Blue Star House, Highgate Hill, London N19
Set in 9D on 11pt Univers
Printed and bound in Holland by
Drukkerij Reclame NV
SBN 289 79718 7 (paperback)
289 79713 6 (hardbound)

Contents

1 How one sees

When all you want is to be told as quickly and as simply as possible where to put the lighting in your home and what to choose from the plethora of bulbs and shades and fittings there are about today, it always seems rather a bore that almost every book on lighting begins by telling you how you see. Brief though the reference to it will be, this volume is no exception, for in spite of all that has been written, said and broadcast on the subject still very few people think of *their* lighting as being connected with seeing in anything but the most perfunctory way; deep down inside they fall back on their primitive cave-man instincts, looking upon it primarily as a means of pushing back the darkness.

If that's all they want it could easily be left at that. But if you feel that there may be rather more to lighting than holding the dark at bay, you will find it far easier to plan it sensibly if you have at least some idea of the fundamental data from which the experts start to design the lighting of a home, and you can only really get this by knowing just a little about how seeing works.

There are three elements involved in the process of seeing - the eye, the object the eye is looking at and the light (either reflected from or given out by the object) - that make it possible for the eye to see. It may help you to appreciate why the eye needs more than just any old sort of light to work at its best if you

Fig 1 Contrast: too great a contrast shows up outline but obscures details.

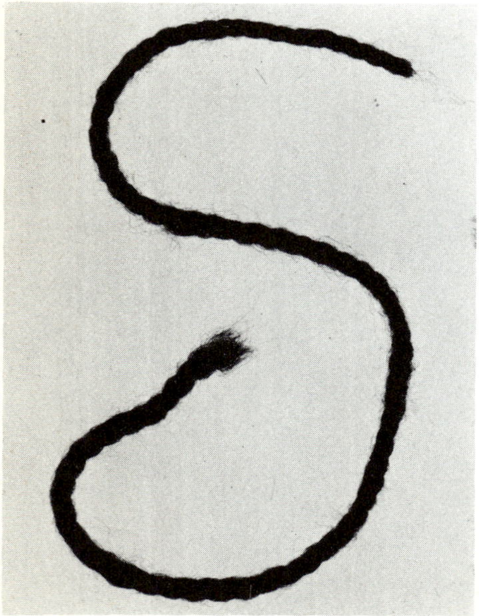

Fig 2 Contrast: all the details can be seen on the same piece of black wool when it is placed on a dark background.

Same lion – different lighting. A startling demonstration of the difference between diffuse and direct lighting.

Fig 3 Diffuse from in front

Fig 4 Direct from above

Fig 5 Direct from both sides

Fig 6 Direct from below

think of it as a camera, for in many ways it behaves like one.

It has a lens; an adjustable aperture - the iris; and, at the back, a film or photographic plate - an area of light sensitive nerve endings called the retina. It is from the retina that messages about what is seen are carried to the brain. How clear these messages are depends in the first place on the condition of the eye; assuming that it is working correctly and that any faults have been compensated for by properly prescribed spectacles or contact lenses, the quality of the 'photograph' or image on the retina depends, just as in a camera, on the quality and quantity of light.

The eye hates having to cope with glare, that is any over-bright light shining directly into it. It finds difficulty in taking clear 'photographs' if contrasts are too low and will also lose some of the detail if they are too high. It will also find it very hard to take good pictures quickly if there is not enough light, and finally, to appreciate modelling and texture, it needs a certain amount of directional light. Give it no glare, reasonable contrast and enough light from the right direction and it will work quickly and easily.

But remember, when comparing the eye with a camera, that is it more adaptable than any camera ever made and in that respect it is probably its own worst enemy; for if it is given time to adapt itself it will still work in conditions so poor that it would be impossible for a normal camera to function. But that doesn't mean to say that it likes them, and while everyone knows that books and newspapers were read, and exquisitely detailed embroidery was done in the days of candles, it does not follow that that is all the light we need now or, indeed, that our ancestors needed then.

Another pitfall for the unwary is to equate the wattage of bulbs and tubes with quality and quantity of light. Some people, for whom aesthetics mean little, will hang one powerful, badly-shaded bulb in the centre of the room and feel that since the light is bright the job is done. On the other hand there are those who are terribly conscious of their visual surroundings, reacting violently against two clashing colours, but who clutter their rooms with vast numbers of lighting fittings, perhaps bought as objets d'art, oblivious of the basic needs of seeing.

In both these extremes the eye will, of course, work. In the first instance, to cope with the harsh glaring light, it will close down its iris - the camera aperture - to keep out the glare and, just as in taking a photograph, then have great difficulty in producing any picture of the shadowed parts of the room - this happens so often in kitchens. In the second case the iris will be open wide to let in all available light but the whole mechanism will be working at fever pitch to get any picture at all, with the result that if detailed seeing, like reading or sewing, is being done, there will inevitably be eye strain. What is needed is lighting that will be right for detailed seeing, lighting that creates the desired decorative effect and lighting that at times need be no more than a glowing ambiance - and if you immediately think that the last just means lighting for TV viewing, forget it. It may be ideal but the modern TV screen needs no special lighting; it is bright enough to compete with daylight and it does no harm if the rest of the room is completely unlit. However some light in the room is probably better than none just for safety's sake, and so that you can see when everybody is sound asleep and can turn the thing off and go to bed.

To return to the various lighting needs: while it is perfectly possible to satisfy them all in the same room it is obviously going to be difficult to achieve them all at the same time. One of the first facts to be faced about good lighting is that it must be flexible and must be designed around the functions of a room; if these functions vary then the lighting must be equally variable. In an office or factory a single unalterable system may suffice, but not in our homes.

There is nothing novel in this changing quality of lighting; it is just, in fact, what Nature has always provided - infinitely variable in intensity, colour and direction. Indeed, laboured though it may seem, it may not be so far out to suggest that as good a guide as any to the sort of lighting you like best in your home is to consider what type of lighting, what season, what time of year gives you the most pleasure in Nature.

Recommended Minimum Values of Illumination in Lux

Living rooms
- General 100
- Casual reading 200
- Prolonged reading and writing 400
- Needlework 600

Bedrooms
- General* 50
- Bedhead 200

Kitchens - work surfaces 200
Bathrooms* 100
Halls and landings 100
Stairs (at tread) 100
Workshop benches 400
Garages - general 50

* Supplementary local lighting should be provided at mirrors

For the elderly, the values given should be increased by 50 to 100 per cent. Particular attention should be paid to avoiding glare and to revealing steps and obstructions.

Measuring Light
A special instrument called a lightmeter can be used to measure illumination values but as a quick ready-reckoner arrange a bare unshaded pearl bulb at the following distances above a book. With no other light present the illumination on the book will be of the order shown below.

100 watts at 2′ give 250 lux
100 watts at 3′ give 100/120 lux
100 watts at 4′ give 60/70 lux
150 watts at 2′ give 400 lux
150 watts at 3′ give 150/200 lux
150 watts at 4′ give 100 lux
150 watts at 5′ give 60 lux

Whatever the part of the world, each time of year has its own particular quality of light and Nature has provided the perfect décor to go with it. But far too many just accept these things without making the effort to appreciate what they are looking at, rather as one might glance at a clock to tell the time while being quite oblivious to what the clock itself looks like. Or, if they do consciously 'see', few people realize how easy it is to re-create the natural lighting effects and colour schemes that they love, in their own homes. If you feel that this is an impossible idea, just think of the shop window displays that follow the seasons every year.

But the most obvious aspect of natural lighting is, of course, what comes in through the windows, and the majority of people feel that this is in no way connected with artificial lighting. This is a great mistake, for the room and its contents remain constant and there should be a constant relationship between all the lighting and all the contents all the time. For example, should the furniture in a sitting-room either be arranged only in relation to the light from the windows or solely to that of a small number of light sources, the room then is really either a 'day room' or a 'night room' but never a room which strikes a happy balance between the two. Present-day thinking is turning more and more away from classing artificial lighting just as an after-dark affair but recognizing that it has a really functional part to play during the day as well.

Consider a fairly deep sitting room with windows on one side only. If the room gets the sun there will, of course, be days in summer when it will look its very best with the sunlight streaming in through wide-open windows or through partially closed Venetian blinds, but there will be many other days when daylight will do no more than make the windows uninteresting bright patches and the areas just inside them the only adequately lit parts of the room. Of course, in such cases, the frankly artificial light can be turned on but then there

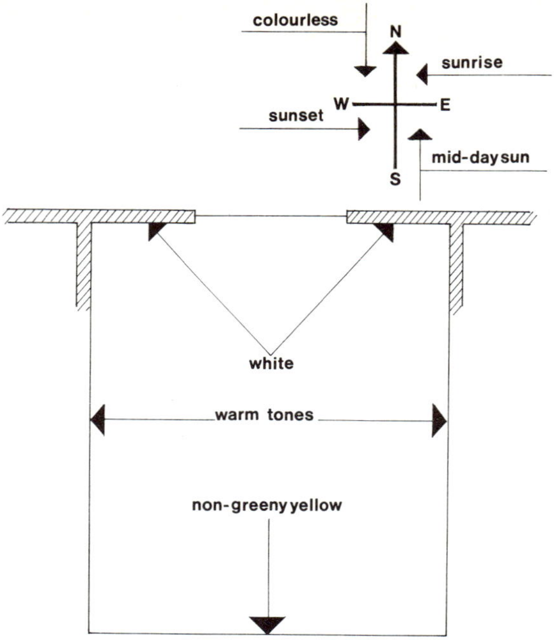

Fig 7 To make a north-facing room feel sunny, apply colours which imitate the basic daylight effects of one which faces south. If the window wall is white, more light will appear to be coming from that direction.

Opposite
Fig 8 The dining end of a deep living room. During the day directional inset downlighters increase the light on the back wall, balancing the brightness of the facing windows. A rise-and-fall appliance is contained in the ovoid on the pendant suspension and the serving hatch has fluorescent tubes above diffusing panels at the top as well as under each shelf.

is not only the rather unsatisfactory psychological effect of an artificially lit room during the day but disparity between the intensity of the level of lighting just inside the windows and that in the rest of the room. It is here that the daytime artificial lighting comes in, for by tactfully flooding the walls with light from concealed fluorescent tubes or possibly by using inset downlighters to increase the light on the floor and furniture farthest from the windows, a balance can be created between the daylight and the supplementary lighting to produce a light room with a daytime feeling to it, and the obviously artificial lighting can be kept for use after dark when the daytime system can be turned off.

Daytime artificial lighting can also be considered from the point of view of economy in building, for in many cases it is considerably cheaper to construct deep buildings where kitchens and bathrooms form the core and have no natural lighting at all. That such rooms ought to be provided with suitable lighting as an integral part of their very structure goes without saying, but to assume that this is always done would be highly optimistic, and people who are used to living with daylight in every room may panic at the thought of surviving with windowless interiors. But this is prejudice rather than reason. With the correct use of light and colour even a no-daylight sitting room could become a perfectly practical proposition, and from the common-sense point of view it may be better to do without the view in rooms where there is little time to appreciate it, if it means lower building costs and the availability of larger areas of outside wall for windows in those rooms that really require them.

But the influence of light on space applies to exteriors just as much as interiors. Not to be able to use a balcony, a loggia or a garden for no other reason than that it is dark is as absurd as to think that traffic on the roads should stop with darkness, and when planning any home lighting installation provision should always be made for the surrounding area as well, even if it is no more than a balcony or window box.

To sum up: good lighting makes seeing easier, it can help create atmosphere and it can help to make the maximum use of space.

2 Bulbs, tubes, types of fittings, caps

Two main sources of artificial light are used for home lighting - filament bulbs and fluorescent tubes but, while the first are completely accepted, quite a number of people still find the second something of a mystery, even though there is hardly an office or factory in which they are not used and more and more are being successfully applied in the home.

Very basically the differences between them are that in a filament bulb the light is produced by electricity straight from the mains heating a very thin tightly coiled tungsten wire until it is whiter than white hot, while a fluorescent tube has no filament and control gear has to be let into the circuit before the tube can be connected.

In fluorescent tubes there is a small coil of wire at each end and electricity jumps backwards and forwards between them to produce ultra violet radiation inside the tube. The inside of the tube is coated with a mixture of fluorescent powders - hence fluorescent tube - and these have the property of transforming this almost invisible ultra violet into visible light of various wavelengths; it depends on the mixture of powders what effect on colour the light from any particular variety of tube will have.

There is no need to be alarmed at the mention of ultra-violet because this can only get outside the glass tube in such infinitesimal amounts as to be completely harmless to any normal person, that is to say anyone who is unaffected by the ultra violet always present in ordinary non-sunny daylight.

The reason that control gear must always be let into the circuit is that in the first place electricity has to be encouraged to jump from one end of the tube to the other - in the case of an 8ft tube this is quite a distance - so it needs something to give it an initial push. But one jump is not enough for the whole time the tube is switched on; the current has to continue to jump backwards and forwards and once it has started it must have something to control it if it is not to be carried away by its own enthusiasm.

However this need rarely concern you for today it is seldom necessary to use anything but complete fluorescent fittings containing all the gear, the old fashioned method of having to place it some distance from the tube - for instance, for curtain lighting where the fitting used to be too clumsy for hiding under the pelmet - only being used in exceptional circumstances.

Having established the basic differences between filament bulbs and fluorescent tubes the following, in very broad terms, are some of the types of both that you are most likely to use.

Filament bulbs

To many people a bulb is just a bulb. They rarely know the wattage they need or the voltage, and as for the type, they will most likely accept anything they are offered. This is roughly equivalent to going to a butcher and simply saying 'I want some meat'.

With bulbs you need to know the wattage because it affects the amount of light that they will give and the amount of electricity that they will use, the voltage because it affects the time a bulb will last; finally, you need to know the type to get the lighting effect you are aiming at.

PEAR (a)
Pearl, clear, white, coloured, rough service, economy nightlight (5 to 8W)

	A		B	
	mm	ir.	mm	in
25 to 100W	60	2 3/8	105	4 1/8
150 & 200W	80	3 1/8	160	6 5/16

THREE-LIGHT (b)

	A		B	
	mm	in	mm	in
40/60/100W	70	2 3/4	120	4 3/4
60/100/160W	80	3 1/8	135	5 5/16

MUSHROOM (c)
White, white with clear crown, pink

	A		B	
	mm	in	mm	in
40W	55	2 3/16	93	3 5/8
60W	55	2 3/16	93	3 5/8
100W	65	2 9/16	109	4 5/16
150W	75	2 15/16	120	4 3/4

'ENLARGED' WHITE (d)
Several types are available; for the one illustrated -

	A		B	
	mm	in	mm	in
40W	70	2 3/4	135	5 5/16

Since effect is so important it may be as well to start with types, for there are hundreds. Fortunately, however, for most straightforward lighting of homes you need only concern yourself in the first place with a basic three, or, in catalogue language, General Lighting Service (GLS) bulbs - pearl, white and clear. The world over they are all made in the familar pear shape but the white is also produced with a mushroom-shaped bulb.

If in doubt, go for the pearl bulb. It gives exactly the same amount of light as a clear one, but with far softer shadows, and it will do for almost any shade or fitting. Should you want a more diffuse - even more shadowless - light, choose the white bulb. It has an inside coating of white powder which makes the whole thing look evenly bright and if you are using a fairly thin shade material it will prevent you from seeing the filament as a bright spot.

As far as the mushroom-shaped version of this bulb is concerned, think before you buy. For a given wattage it may be smaller than its pear-shaped counterpart and consequently where you want more light, possibly from a table lamp, it may seem a good idea to substitute a 150 watt mushroom for a 100 watt pear. If you have a nice, large, airy shade then it is certainly a good idea, but if there is any question of having to 'squeeze' the bulb in to any fairly enclosed shade or fitting, don't use this type, for although it may be no larger than the original 100 watt bulb it will produce more heat, and while you're unlikely to cause a fire you might very well ruin an expensive shade through scorching. Another small point: if you are increasing the wattage in a table lamp in order to get more light downwards, it would

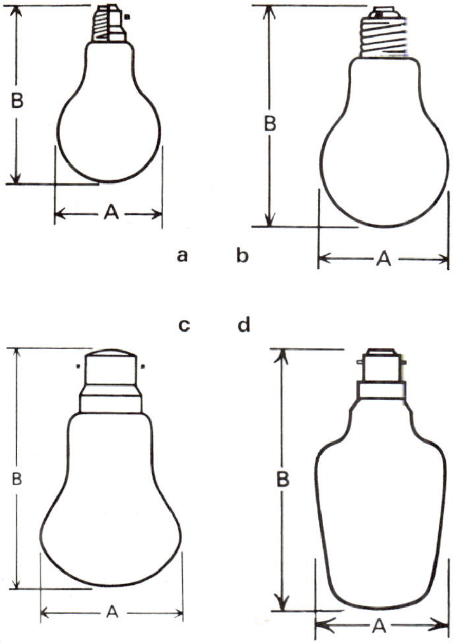

be better to use a pear-shaped bulb since the mushroom type pushes rather more of its light through the end opposite the cap. Incidentally, bulbs last longer if they are used with the metal part - the cap - upwards. Since in nearly all portables they have to be used cap down this may seem slightly devious practice on the manufacturers' part. It isn't. They put vast sums of money into research and if it were an economic proposition to make a filament bulb which lasted longer in an upside-down or sideways position they would make it.

Before leaving white bulbs, there are three others that should be mentioned. One is the three-light bulb containing a 60 watt and a 100 watt filament. Used with a switched lampholder which can bring into circuit either one of the filaments or both, the volume of light is controllable in three steps to suit the purpose required. This is a good idea for table lamps which at times need to do no more than look alight and at others are required to give a really good light; but it has a special cap and requires a special holder which may not be easy to obtain. Holders with adaptors that fit into ordinary filament bulb holders are available but are rarely stocked by retailers. They are worth running to earth if you like the idea.

The second bulb is very much enlarged and is a diffusing fitting in itself. Some manufacturers make special equipment for it but it is a useful one to bear in mind for those situations where you might be tempted to use an unshaded ordinary pearl or white bulb. It costs more than GLS bulbs but lasts longer.

The third is a white bulb with a clear patch left on its crown. In the pendant position this gives nearly as much upward light as an ordinary white bulb but more downward light; it would be suitable, say, over a dining table or in a pendant bedhead fitting.

The last of these GLS bulbs is the clear one. This has been much maligned and indeed at one time, when lighting engineers became positively besotted with abolishing shadows, the domestic sizes looked like being wiped out altogether, which would have been a great pity for there are places where its hard, sharp light is a distinct advantage. For example, a clear bulb in an open-topped opaque shade on a table lamp placed beside a vase of tall grasses or twigs will produce good shadow effects on walls and ceilings and, of course, clear bulbs are nearly always essential for making the most of fittings which are designed to sparkle; there are many delightful modern designs of these which are far removed from the crystal chandeliers and wall-brackets which usually spring to mind when sparkle is mentioned. Never the less be wary of them, for they can also create hard dark shadows where you want them least, say, in kitchens and on the stairs, moreover, they are distinctly unflattering; a clear bulb overhead will dutifully show up every wrinkle in sharp relief.

You will find a definition of voltage on page 77 but we shall now consider why it is essential that you know what the voltage of your mains is.

All bulbs are designed to give a certain amount of light and work, on average, for a certain number of hours. These things have been arrived at very carefully and, taking everything into consideration, the most economic design life for GLS bulbs has for many years been accepted as 1000 hours. Supposing your supply is at 240 volts, and you are sold a bulb designed for 230 volts, while it will give more light, it will only last about half as long as a bulb of the correct voltage rating. If you do it the other way round, while the bulb will last longer, you will get less light than you had hoped for. Although some bulbs have a 2000 hour life, they cost more and give rather less light. They might have a significant advantage if you had to pay someone to renew a bulb when it failed, but when you do it yourself you don't charge for your own time. All the same, they are time-savers in awkward positions and in exterior fittings which need unscrewing to get at the bulb.

And finally wattage, which is not just a mysterious figure on the end of the bulb. It is again someting you should know about. Most good quality fittings, to which overheating can cause damage, are marked with the recommended bulb wattage, and this you should treat seriously, for it matters. Where you will have to use your own judgement is over floor and table lamps. As long as they have orthodox shades, open at the top and bottom and with plenty of clearance round the bulb, a simple guide is to use 100 watts for table lamps and 150 watts for floor standards. When you are shopping, do try to get out of the habit of buying just one bulb at a time. Every house ought to have a bulb store which is as regularly restocked as any other store cupboard, for to find at some vital moment that there is no replacement or only a bulb of the wrong type and wattage for some critical position is not only inconvenient, it could be positively dangerous.

Filament bulbs, of course, don't end there. There are several other types that can be used and it is worth knowing of their existence. The following are just a few.

Coloured Yellow, blue, red, pink, amber, green, internally coloured bulbs. Rarely use them as part of the day to day lighting scheme, but keep a few on hand for changing the look of everything for parties. They're also fun for out of doors.

Candle A large variety of shapes and finishes. Absolutely necessary for wall brackets with small slim shades and with most branched pendants, chandeliers and candelabra. Some even have a built-in flicker.

Tubular Most people today on hearing 'tubular' are inclined to think at once of fluorescent tubes, but there are a number of filament tubes

a
b
c

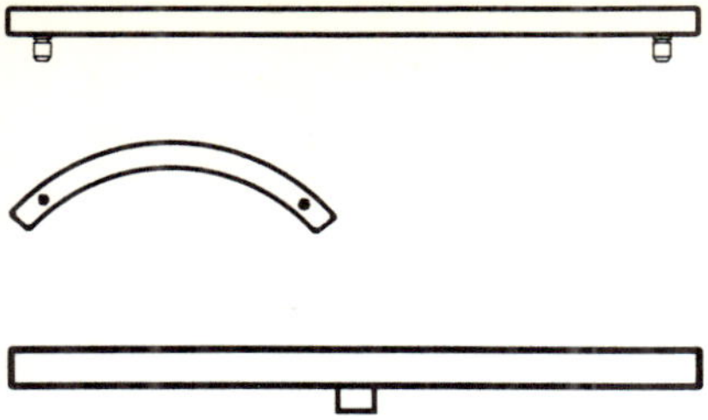

Architectural

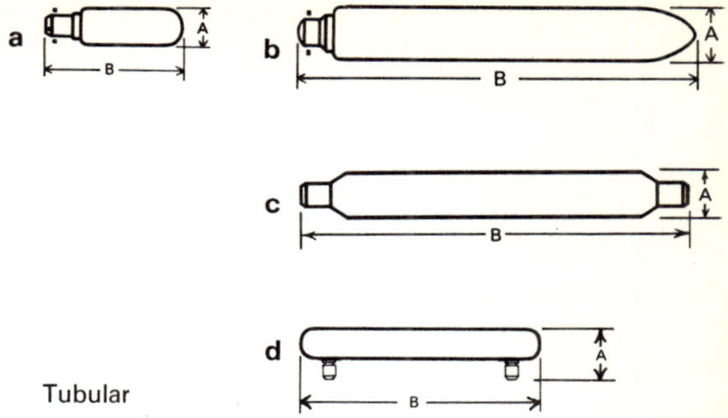

Tubular

Architectural

STRAIGHT, WITH TWO PEG CAPS (a)
White

	Length		Diameter	
	mm	in	mm	in
35W	305	12	30	$1\frac{3}{16}$
53W	457	18	30	$1\frac{3}{16}$
60W	500	20	30	$1\frac{3}{16}$
75W	610	24	30	$1\frac{3}{16}$
110W	915	36	30	$1\frac{3}{16}$
150W	1220	48	30	$1\frac{3}{16}$

CURVED, LENGTH 500mm (20in) (b)
White
60W $\frac{1}{2}$ circle of radius 159mm ($6\frac{1}{4}$in)
60W $\frac{1}{4}$ circle of radius 318mm ($12\frac{1}{2}$in)
60W $\frac{1}{8}$ circle of radius 636mm (25in)

STRAIGHT, WITH SINGLE CENTRE CAP (c)
White

	Length		Diameter	
	mm	in	mm	in
60W	500	20	30	$1\frac{3}{16}$

Tubular

SINGLE CAP (a)
Clear or frosted

	A		B	
	mm	in	mm	in
25W, 40W	25	1	82	$3\frac{1}{4}$

SINGLE CAP (b)
Opal

	A		B	
	mm	in	mm	in
60W	37	$1\frac{7}{16}$	302	$11\frac{7}{8}$

DOUBLE CAP (c)
Clear, opalized, pink, amber

	A		B	
	mm	in	mm	in
30W, 60W	25	1	221	$8\frac{11}{16}$
30W, 60W	25	1	284	$11\frac{3}{16}$

DOUBLE CAP (d)
Clear, opalized

	A		B	
	mm	in	mm	in
40W, 60W	25	1	252	10

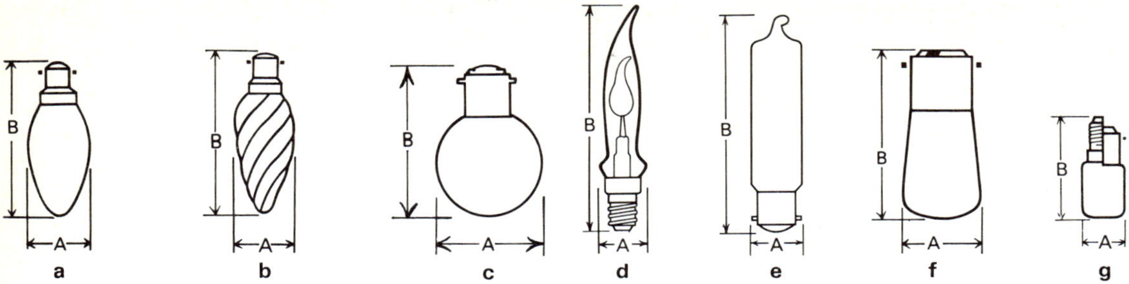

Candle and miscellaneous

PLAIN (a)
Clear, frosted, white, coloured

	A		B	
	mm	in	mm	in
25W, 40W	35	1 3/8	92	3 5/8
60W	45	1 3/4	123	4 7/8

TWISTED (b)
Clear, frosted, coloured

	A		B	
	mm	in	mm	in
25W	35	1 3/8	95	3 3/4
40W, 60W	46	1 3/4	123	4 7/8

ROUND BULB (c)
White

	A		B	
	mm	in	mm	in
24W, 40W	45	1 3/4	65	2 9/16

'FLICKER FLAME' (d)
For flicker effect only, very little light

	A		B	
	mm	in	mm	in
3W	16	5/8	80	3 1/8

'PICKWICK' (e)
White

	A		B	
	mm	in	mm	in
40W	33	1 1/4	128	5

NEON INDICATOR (f)
Clear

	A		B	
	mm	in	mm	in
0.5W	28	1 1/8	56	2 1/4

PILOT (g)
Clear

	A		B	
	mm	in	mm	in
6W, 10W	19	3/4	48	1 7/8

Fig 9 Unique, and the nearest thing to a real candle. The 'flame' is a separate low-voltage 1.75 watt bulb with a 'micro' screw cap and the candles come in 6", 9" and 12" lengths either tapered or straight-sided. They are very expensive and need transformers which can be placed up to forty feet away, but for a really good chandelier there is nothing better.

which will get light into spaces which would be much too narrow for ordinary bulbs. They give far less light than fluorescent tubes but they do not require control gear, although they do need special holders.

Reflector These bulbs are partially mirrored inside so that they have their own built-in reflectors. Some are designed to give narrow beams for spot lighting and others have wide beam spreads for flood lighting. The PAR bulb, a tough version that can be used unprotected out of doors, also comes in red, yellow, blue and green and the latest to be added to the family are 'dichroics'. These have the remark-

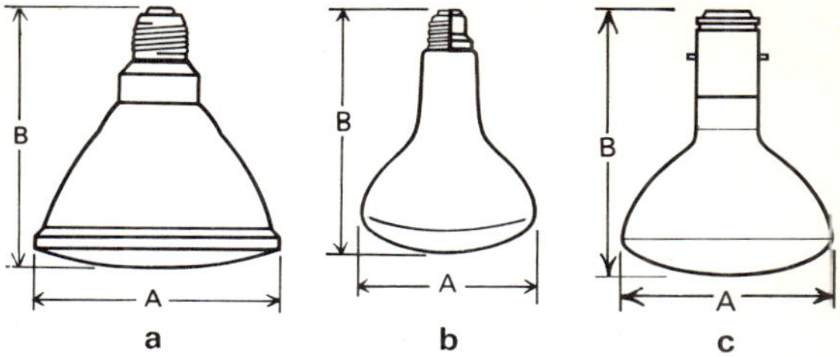

Reflector

PAR BULB (a)
Spotlight or floodlight. Plain or coloured, also dichroic

	A		B	
	mm	in	mm	in
150W	121	$4\frac{3}{4}$	135	$5\frac{5}{16}$

BLOWN BULB (b)
Spotlight or floodlight

	A		B	
	mm	in	mm	in
100W	95	$3\frac{3}{4}$	135	$5\frac{5}{16}$
150W	126	5	178	7
250W	126	5	178	7

LOW VOLTAGE (c)
12 volt, 50 watt. Representative of low voltage reflector bulbs

	A		B	
	mm	in	mm	in
	64	$2\frac{1}{2}$	82	$3\frac{1}{4}$

able character of not projecting heat in the same direction as the light, thus in situations where you don't mind heat above the ceiling or in the upper part of the room, they're good for giving a cool concentrated light below, for example, where you need spotlighting in the kitchen. Besides this they can be obtained in coloured versions which give a far purer coloured light than can be obtained by any other means. Two or three different colours used at the same time in separate adjustable fittings are marvellous for creating multi-coloured shadows.

Spotlights Different from the reflector bulbs, these are much smaller and give more concentrated beams. Some, because they require a voltage lower than that of the mains supply, need transformers, and each is designed for a special type of fitting or reflector. They can be very effective if used correctly.

Christmas Tree The fascinating variety of these bulbs, particularly the miniatures, should not be kept just for Christmas. A set in a single colour attached to twigs or let in to perforations in a wood or metal shape can be part of the permanent décor and they look delightful arranged at random behind a translucent drape.

Fluorescent tubes

Fluorescent tubes last several times as long as filament bulbs and give several times as much light for a given wattage, but both these statements need qualification.

The nominal life of the shorter tubes is 5000 working hours; that of most tubes 4' long and over, 7500 hours - rather more than six hours' light every day for three years. Frequent switching on and off tends to shorten the life, but unless you are one of those people who makes a fetish of switching off there is really no need to worry about this.

The efficacy of fluorescent tubes - that is, the amount of light they give for the current consumed - depends both on their wattage and nominal colour. Taking for example the 40 watt (4') the efficacy is about 66 lumens per watt - about seven times that of a 40 watt filament bulb; in other words, one would need seven 40 watt bulbs to give as much light as one 40 watt tube, although this does not mean that a simple substitution of tube for bulbs will always suffice, for the light from the tube will be coming from only one position whereas the bulbs can be scattered around the room.

This very high efficacy is for the White tube colour. There are other colours available, some giving a warmer light, some colder, some giving a more pleasing effect to decorations and so on. Unfortunately the tubes with the highest efficacy do not have the most pleasing colour effect, and the better colour tubes, called the De luxe varieties, in general give about one third less light. Thus the choice of tube should always be made first on a basis of whether efficacy - amount of light - or pleasant colour is the more important for the particular application. In homes, pleasant colour should usually come first, but with the right decorations tubes in the high efficacy group can be quite acceptable.

You may find, however, a White or Warm White 4' 40 watt tube packed with a lighting fitting designed for kitchen use. This happens because manufacturers are anxious to provide the maximum amount of light for your money. Since every colour of tube is interchangeable with any other you can usually have the tube changed at no extra cost if you do not like the effect. The important thing, however, is not to think that all tubes produce the same colour of light if such a packaged

Light output and colour effects of fluorescent tubes

Tube	Relative light output	Colour effect: Décor	Colour effect: Complexions
White	100	Emhasizes yellow and green. Poor on red.	Unkind
Warm White	100	Emphasizes yellow. Poor on red.	Unkind
Daylight	95	Emphazises yellow and green.	Unkind
Natural	75	Like sunny daylight.	Kind
De luxe Warm White	65	Like filament bulbs.	Kind
Colour Matching and Northlight	65	Like north sky daylight.	Unkind but truthful
De luxe Natural	55	Emphasizes all colours, particularly red.	Kind

Names of tubes vary between manufacturers; those given are the best known in the UK. For comparison, the relative light output of a filament bulb of the same wattage as the fluorescent tube is about 25.

Symptoms of faults in fluorescent tubes

Towards the end of its useful life a tube will become blackened for up to 2" from both ends. This is due to evaporation of electrode material, and usually cannot be prevented. With new tubes, a swirling effect as if the arc were spiralling within the tube can sometimes be seen. The swirling should disappear after a short while, and this is usually helped by switching the tube on and off a few times. If the effect persists, try a new starter switch. Faults in the tube or circuit can often be diagnosed by observing the tube's behaviour, thus -

Symptom	Probable cause
Tube appears completely dead.	No mains volts. Faulty or inappropriate starter. Break in circuit; suspect blown fuse inside fitting, or (switch-start) broken tube electrode or faulty lampholder contact
Tube glows continuously at both ends but makes no effort to start.	(Switch-start) Faulty starter. If end glow is red, not white, tube has passed useful life (Starterless) Ineffective earth connexion to fitting or to metal stripe on tube
Tube glows at one end when trying unsuccessfully to start.	(Switch-start) Lampholder connexions at dead end of tube short-circuited. (Starterless) Broken tube electrode or disconnexion to dead end of tube
Tube appears normal but makes repeated unsuccessful efforts to start.	Faulty starter. Very low mains voltage. Very old tube

Tube lights normally for a few seconds then goes out, and repeats.	Faulty starter or very old tube. Very low mains voltage
As above but extinguishing with a shimmering effect.	Very old tube
Tube lights up at half brightness, with obvious flicker.	Very old tube

With a starterless circuit it will normally do the circuit no harm to leave a faulty tube in place, but with a switch-start circuit a faulty tube left for a long period before being replaced may cause the control gear to overheat and/or the starter switch to fail. This may be avoided by using cut-out starters.

fitting should be your first experience of fluorescent lighting.

There are, too, smaller fluorescent tubes for which there are many applications in the home - the 4 watt, 6 watt and 8 watt. For example, for night safety lighting on stairs, a 6 watt fluorescent tube would give enough light in an otherwise dark house - rather more than a 25 watt filament bulb - and is useful, too, in exterior fittings.

The day by day consumption of electricity by tubes is unchanged throughout life, but the light output falls slowly with time even if you keep them clean. At the end of its nominal life the light output is some 20 per cent lower than when the tube was new. It is worth making a note of when tubes are first put in, for most last far beyond their nominal lives and the lighting can easily get duller and duller without your noticing.

It has already been said that every fluorescent tube must have control gear in its circuit to help start the tube and to control the current to the correct value. This normally takes the form of a choke or transformer and associated capacitor ready-wired in the fitting, but the components can be obtained separately where necessary. A starter-switch, which is a replaceable item, is an additional component with some circuits. In general, circuits without starter-switches require an earth connexion to ensure that the tube will start reliably. Economy in initial cost is in a few cases obtained by using a simple resistor in place of the choke and capacitor, but the circuit then consumes about twice as much electricity and is therefore unlikely to be economic in the long run.

Fittings

It is often not quite clear what is meant when people talk about a 'lighting fitting'. It is a term used to describe any ready-made piece of equipment for use with a bulb or tube other than an unadorned lampholder.

Among the main types of fitting are:

Pendants Hanging fittings, which are supported by chains, rods, cables, or, if they are light enough in weight, the wire or flex which carries the current.

Inset Any fitting which is partially or wholly set into a ceiling. Apart from some forms of night-lighting which are referred to later as 'after lights out lighting', one very rarely comes across fittings set into walls.

Ceiling or flush mounted This usually refers to any fitting which is closely attached to the ceiling, that is, without the suspensions that pendants have. And in this context a definition of the term 'mounted' may be helpful, since lighting engineers have invented their own meaning for it. The simple interpretation is, 'attached to the surface of'.

Wall brackets and wall mounted Anything from designs based on the very earliest torch and lantern holders to the most up-to-date fluorescent fittings.

Portables Mainly floor and table lamps, sometimes called floor and table standards, but anything that is plug-connected and can be easily moved. You can have portable wall-brackets that just need one slender pin to keep them up, portable spotlights, portable floodlights - in and out of doors - and if you have hooks in your ceiling there's no reason why you shouldn't have portable pendants too.

A further step that may help to clarify what follows is to classify lighting fittings in terms of how they distribute their light. A fitting giving all or very nearly all of its light downward is called 'direct', and all upward, 'indirect', while a general diffusing fitting gives fairly even lighting in all directions - a typical example is the white glass sphere so frequently found in bathrooms and kitchens. The intermediate types are called 'semi-direct' and 'semi-indirect'.

Caps

If you were on the receiving end of enquiries to bulb manufacturers you would very soon realize that many people when buying a complete fitting never think about the question of whether the bulb is going to be difficult to replace, for they do not know that besides the wide variety of bulbs, there are several alternative types of cap, i.e. the metal parts. When buying fittings - especially imported ones - find out if replacements are easily available if the cap appears to be at all unusual. If you do need a new bulb, always take the old one with you to make sure of getting the right cap to fit the holder. The following will give you an idea of the variety.

Bayonet Cap (*BC*) The standard cap in the UK for all ordinary GLS pear and mushroom shaped bulbs up to 150 watts, for switchboard and indicator bulbs and one of several alternatives for candle and decoration bulbs.

Small Bayonet Cap (*SBC*) *and Small Edison Screw* (*SES*) Mainly alternative caps for candle and decoration bulbs.

Goliath Edison Screw (*GES*) For GLS bulbs of 300 watts and above.

Edison Screw (*ES*) For 100 to 250 watt reflector floodlights and spotlights, and alternative to BC for 150 watt GLS bulbs.

Miniature Edison Screw (*MES*) *and Candelabra* Mainly for very small bulbs such as the Christmas tree type.

Single Centre Contact (*SCC*) For filament tubes with a cap at each end. Architectural tubular lamps are fitted with special peg caps.

Bi-pin For fluorescent tubes of 15 watts and above

Miniature bi-pin For fluorescent tubes below 15 watts.

The cost of lighting

Few people realize how little lighting costs to run. The tables show how many hours of light are obtained from popular sizes of bulbs and tubes for one unit of electricity. On average, the 'good colour' or De luxe types of fluorescent tube give about four times, and the high efficacy types about six times, as much light as filament bulbs of corresponding wattage.

Filament bulbs All shapes and types	Hours
25 watts	40
30 watts	33
40 watts	25
60 watts	17
100 watts	10
150 watts	7
200 watts	5

Fluorescent tubes All colours	Nominal length in	 m	Hours
8 watts	12	0.3	67
13 watts	21	0.5	48
15 watts	18	0.45	25*
20 watts	24	0.6	20*
30 watts	36	0.9	25
40 watts	48	1.2	20
50 watts	60	1.5	14
65 watts	60	1.5	12
85 watts	72	1.8	10
85 watts	96	2.4	10

* per pair of tubes

3 Planning a lighting installation

There is no single 'right way' to light a home. Lighting should suit *your* convenience and serve *your* habits and as long as you are strong-minded enough to reach a conclusion at some point, you cannot give too much thought to planning it. The following questions are intended to help you avoid that agonizing situation of having looked at what seems like every lighting fitting under the sun and still not having the faintest idea what to buy. If the answer in every case is 'Yes', then you should be well on the way to reaching a solution. If it is 'I don't know', then it's difficult to know what to suggest, for even if you called in the best lighting specialist in the world it would be almost impossible for him to reach any conclusion either, for lighting is a functional thing and unless you know the functions you want it to perform you've got no starting point. This is one of the things that makes it so difficult for architects to provide the right number of outlets in the right positions for lighting in large housing projects. They can do no more than make intelligent guesses based on the assumption of average needs and it is quite wrong to judge their capacity for imagination and ingenuity on such schemes.

If they knew the answers to these questions for every household they would almost certainly, in many circumstances, produce very different results, so take your architect into your confidence for you will find that he is as keen as you are to see that your lighting does everything that you want it to do.

But whatever happens don't let yourself develop an inferiority complex from what you have seen in magazines and exhibitions. And do not let anyone lead you to believe that lighting can be settled in a quick wander around an empty house or flat or that it needs no more than an hour or so making marks on a plan. Cost apart, it doesn't work like that. You must have time to think and, when you've thought, time to relate your thoughts to the equipment available and, if the household consists of more than you, time for discussion.

Assuming, then, that you're starting from scratch - that is to say, with empty existing rooms or rooms that are still lines on an architect's drawing - consider these questions.

1) Do you know what each room is going to be?

Not long ago it was considered a bit of a let down if a house did not face south, at least in England. With small houses this usually meant that the master bedroom and the dining room got the best of the mid-day sun, which scarcely made sense. So if it isn't the sort of modern house where all the thinking has been done for you, don't be afraid to change things. An upstairs sitting room facing west might be a very much better idea than using the room for a bedroom which will be unoccupied for most of the time when the light is at its best, even though it may mean more stair-climbing and giving rather more thought to the landing décor.

2) Do you know how long you are going to live in the place?

This may easily influence the answer to the first question. Your ideas may change beyond recognition between, say, the twenties and forties, and even the forties and fifties, and if you have a young family remember that it grows up faster than seems possible. Besides

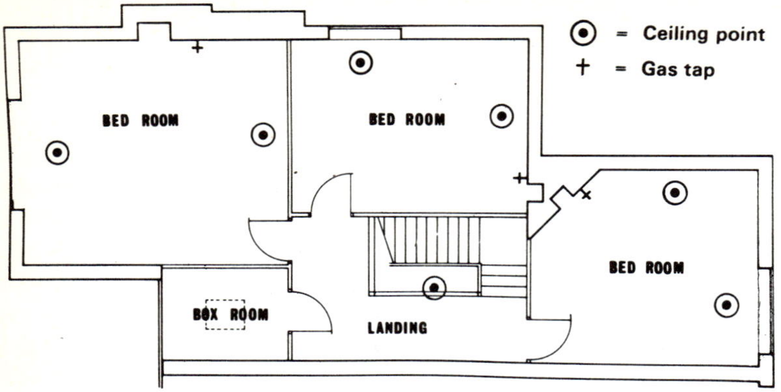

Fig 10 The lighting requirements in the conversion of, say, the top floor in a three-storey Victorian or Edwardian house to an all-electric self-contained flat.

Original with ceiling points only and gas for fires.

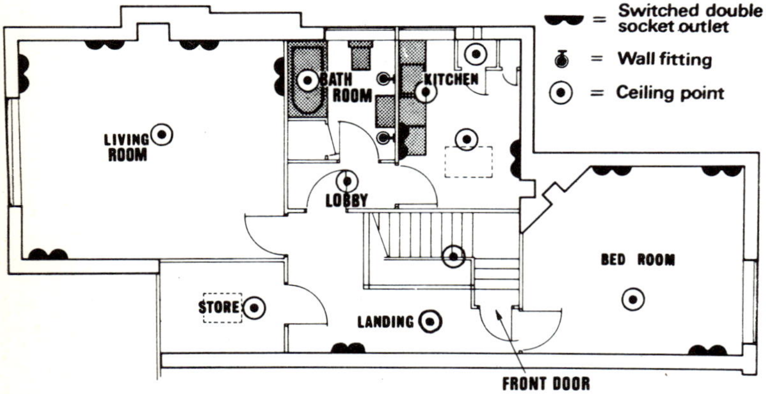

After conversion there is adequate provision for good lighting and other electrical appliances.

this, from the age of about forty onwards, most people's eyesight gradually deteriorates; they need more light as time passes, and it is sensible to make provision for this at the start, thus avoiding costly subsequent additions.

3) Do you know how the rooms are going to be used?

This doesn't just mean as a sitting room or kitchen or dining room, but covers how many people are going to use a room at any one time, and the sort of things they're going to want to do in it.

4) Do you know - really know - whatfurniture is going into the rooms?

5) Do you know how the furniture is going to be arranged?

Except in very small rooms where there is no space for shifting things around there may be many practical answers, and one way to avoid making expensive mistakes with wiring for lighting is to draw a plan on squared paper in feet or metres or any unit you like, as long as it's accurately to scale, and cut out the shapes of your furniture and shuffle them around. Don't forget, though, to mark the height on each piece; many of those crowded exhibition rooms one sees are caused by people forgetting to do this. But better than the simple plan, if you have the energy and skill, is to make a genuine three-dimensional model. More and more exhibition and stage designers are doing this today and it is eminently worth while. The simplest representative furniture will do, but if you want to go further buy toy furniture which is often made to a definite scale, and work backwards from that.

Whichever method you use, try every variation you can think of, taking into consideration the view from the window, the daylight, the heating, winter and summer, and, if you're using a model, put a ceiling on and look through the door to see the effect that real daylight gives; and don't forget mirrors and pictures, for they may well need special lighting before you've finished. Also it's not a bad idea to measure the girth of the fattest person in the house and turn them into a bright red disc or tube to make sure that you've left enough space for movement. When you've reached the best compromise - you've really no right to expect more - stick things down lightly and start to think of the answer to the next question, although it will not be dealt with in detail until later (Chapter 5).

6) Do you know what colours and textures you're going to use?

At this stage you may well be wondering what all this has to do with lighting, and if you've never considered lighting as an integral part of your decoration and furnishing, indeed, of your entire way of life, your bewilderment is not surprising. But these are all things that should be known in order to plan a good lastingly useful lighting installation. If you have lived in a place for years and can firmly answer 'Yes' to everything and still have not got good lighting, then you should prepare yourself for quite a major rewiring job if you really want an improvement. One of the main points of asking all these questions is to save anyone moving into a new house the upheaval and expense that rewiring necessitates. A sound wiring installation with enough outlets where you can connect lighting equipment is the foundation of any good lighting scheme.

Wiring

Part of almost all new wiring today is done on the ring circuit method. This is just what it sounds like; a ring of wiring running around each floor or flat to which are connected socket outlets to take plugs which contain their own fuses, 13 amp for heating, 3 amp for lighting. This ring can run under the floor, in the wall, above the ceiling, anywhere, in fact, that it is out of sight; but since one of its advantages is that an unlimited number of socket outlets may be connected to any one ring, provided it does not serve an area of more than 1000 square feet, it seems a great pity that many homes have ring circuits buried in the structure, so that if you want to add to them you've got to dig up the walls. One of the best methods, though you'll seldom find it done, is to install the ring circuit behind easily detachable skirting boards, making sure that there is plenty of slack cable. In this way if you need to add extra socket outlets at any time the job can be done without a lot of fuss. Dome-headed screws are best for fixing such skirtings, but if you don't like the look of them, ordinary counter-sunk screws will do.

Ring circuits are normally used only for carrying socket outlets, with a separate system

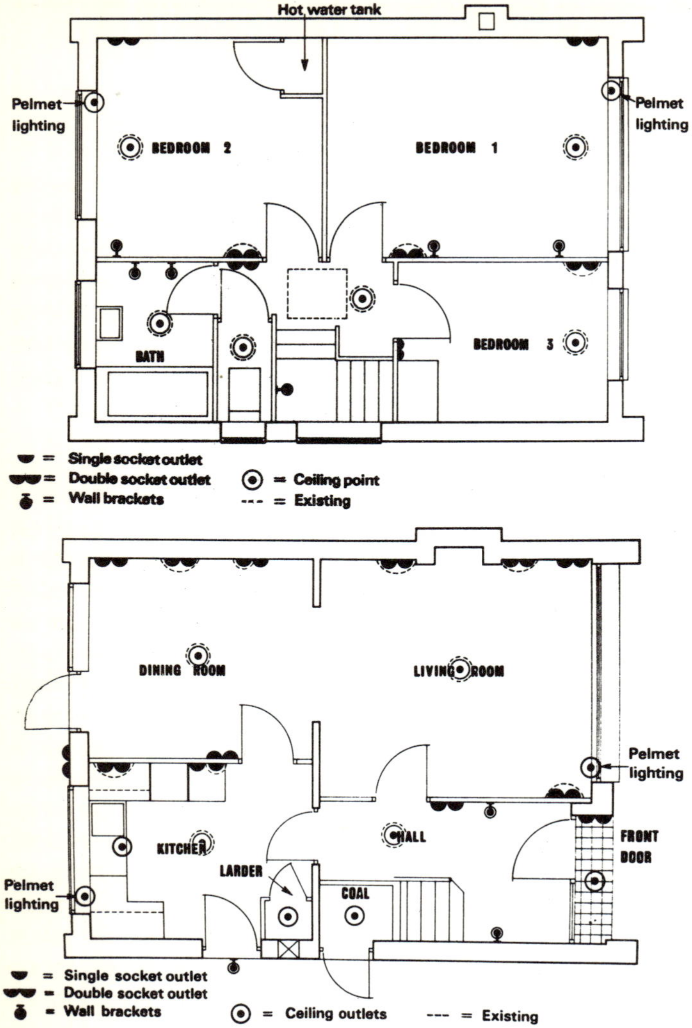

Fig 11 Very few new houses are adequately wired. These ground and first floor plans of a small three-bedroomed house show the builder's original installation and the additions that should be made to bring the electrical installation up to modern standards.

of wiring for ceiling and wall outlets being taken back to a circuit breaker or conventional fuse box, but this is not strictly necessary. The entire lighting installation can be wired from the ring: all it means is that where you need a 'branch' or spur to leave it for feeding a ceiling or wall point the wiring must pass through a fused spur box. In this way you never have to go to a central point to replace a 'blown' fuse. The fuse, if it blows, is on the spot and is even easier to replace than that in the normal plug top. Correctly installed, the fused spur type of installation complies with all known wiring regulations.

Obviously, for complete convenience you need a separate socket outlet for every portable lamp and electrical appliance - the approximate number will emerge in due course - but the position of socket outlets needs consideration. It is easy enough, if you haven't thought about the questions posed earlier and you haven't drawn plans or made models, to make a snap decision to have so many for each room, apparently sensibly distributed around the skirting, and, of course, it may work. On the other hand, you may find that you finally have a heavy chest or piece of built-in furniture in front of a socket outlet. So plan their positions carefully and don't necessarily feel that they must all be in or just above the skirting board. Ideally they ought all to be of the switched type but even if they are not it may be infinitely more convenient to have them installed at a height which avoids stooping or, thinking again of the heavy chest, put in just above the top of the chest so that you can easily plug in a table lamp. And always have double socket outlets rather than single ones. They are only fractionally more expensive and twice as convenient.

While it is a nuisance to have socket outlets of different sizes, in a room lit solely by portables it may be a good idea to install a number of 2 amp outlets which can be on a separately fused circuit controlled by a main switch at the door.

If the house is still at the drawing board stage, or you have embarked on a major rewiring job, make sure that meters and circuit breakers or fuse boxes are easy to get at and well-lit. Very near the front door is the best position or possibly in a downstairs cloakroom; and keep a torch in the cupboard in case there is a failure after dark - although if you make good use of the ring circuit there should be no part of the house which can be plunged into complete darkness. Let us take the hall as an example. If the main ceiling light is on one fuse or fused spur, wall brackets should be on another, or if you have a table lamp on the ring then that will be unaffected if the bulb in the ceiling fitting goes or the fuse blows. The days of nearly all the lighting on a floor being governed by a single fuse in an awkwardly placed box under the stairs should be a thing of the past.

Switching

The obvious place for switches in a room is beside a door but this does not mean that it is always the best. The best place for a switch is where it is most convenient for you to find it and while this certainly means at least one by the door to switch on a light when you come in, it may well be sensible to scatter them around. Pull-cord switches are excellent money savers and should have found their way out of the more utilitarian rooms far more frequently than they have. Kept close to walls and suitably coloured for camouflage they can provide convenient switching for ceiling lighting far more cheaply than normal wall switches, for it is simply a question of a short wiring run above the ceiling from the fitting to the point where you want the switch.

In place of normal wall switches small dimmers can now be installed. All these are suitable for use with ordinary bulbs and some of them will work with fluorescent tubes, provided that the control gear is specially designed to allow dimming. With the ordinary fluorescent control gear it can't be done, but who really wants to dim fluorescent lighting in the home when such wonderfully dramatic effects can be created with filament bulbs? Just to take a small flight of fancy... Imagine you have a charmingly

Fig 12 A typical dimmer switch which can be substituted for an ordinary switch. This one will cope with 300 watts of filament lighting – there are others for higher loads and for fluorescent tubes.

Opposite

Fig 13 An unusual pelmet which has been cut out at the top to allow light to reach the ceiling. Chunky square glass wall-fittings contain coloured bulbs purely for decoration and clear candle bulbs are scattered behind the translucent multicoloured shape. The level of light in this windowless hall can compete with daylight and it is obtained with fluorescent tubes above low-brightness inset louvered ceiling panels.

furnished Regency room with two dimmers built into a neat control panel near your usual chair, one to control the bulbs in a crystal chandelier and matching wall brackets, and the other to control two deep blue spot lights trained on to the chandelier from concealed positions at opposite sides of the room. It is a summer night, the windows are open on to a moonlit garden and while you are talking you gradually dim the crystal fittings and bring up the blue, and magically the chandelier changes from a lighting fitting to a sparkling azure jewel and the ceiling and walls are flecked with coloured refracted light.

It certainly makes a change from thinking of dimmers as pieces of useful ironmongery! Do consider them, for they have many uses and are not expensive or difficult to have installed.

Before leaving switching, just a reminder that two-way switching does not only apply to stairs and halls. It's useful in bedrooms, long passages and in any room which has more than one door.

You may now have some elementary thoughts at the back of your mind about wiring, but don't feel that you're still no nearer getting down to the actual lighting. It's rather like redecorating a room. Putting on the final coat of paint and the final wallpaper is fun - it's all the preparation that is such a bore. In the next two chapters you should begin to see some results; first dealing with the basic needs of each part of a home and then considering how you can elaborate on these.

SHIFT

Fig 14 The open plan living area of a penthouse suite. The stairs to the upper floor are of glass, and each is lit by a separate fluorescent tube concealed in a special fitment attached to the back edge. *Architects: Maurice Meyersohn, Sachs and Partners.*

The word home can mean anything from a one room flatlet to Buckingham Palace or the White House. It's all a matter of scaling things up or down, for the same lighting essentials apply throughout. In order, however, not to leave the following recommendations too much in the air, it can be assumed that they are loosely based on a four-bedroom detached family house with a surrounding garden, although outside lighting will be left until the later chapter on exteriors.

Hall, stairs, landing

Unless you are proposing to create some dramatic change in decoration between the hall and the landing, these areas and the stairs can be considered as a whole and lit by matching types of fittings from a single range. Particularly in smaller houses, using the same scheme of decoration, the same carpet and the same style of lighting fittings throughout, binds the space together and makes it seem larger.

Where there is a staircase well, a cluster of fittings at different heights can be hung from a point on the landing ceiling over the well with possibly the lowest two fittings below the hall ceiling. If you have difficulty in finding a ceiling plate or spreader that is small enough to allow the fittings to hang within the available gap, have a wooden disc of the right size made. If no two fittings are hung at the same height then you have only to allow clearance for the flex supporting them. When providing wiring for, say, a five light cluster, it is an economy to have the top and bottom fittings on a separate switch so that they alone can be used should it be necessary to leave the lights on for late home-comers. Forty watt bulbs should be sufficient, but this depends very much on the size, shape and colour of shade; the maximum recommended by the manufacturer should not be exceeded.

For narrow halls and landings a separate ceiling fitting will be necessary for each, but it is essential that the hall fitting should not appear glaring to anyone coming down stairs. For this reason it may be better to avoid pendants and choose something that is ceiling mounted: this could well be a fitting containing a circular fluorescent tube. More accidents in the home happen on the stairs than anywhere else, and bad lighting definitely contributes to them. If you are not satisfied that the hall and landing fittings light the stairs adequately, a strategically placed wall bracket with a good downward light distribution will often provide the answer. But while it should be high enough not to be knocked, beware of placing it in such a position that cleaning it and changing the bulbs is difficult. Depending on the size and shape of the hall, extra wall brackets may be a good idea. For example, if there is a mirror it is logical to place one at each side, they could also be useful in long passages which should not have to rely solely on spilled lighting from the hall. If you prefer fluorescent fittings, choose a type that has semi-opaque or prismatic sides and mount them across the passage: this will tend to have a widening effect.

Having already mentioned daytime artificial lighting as opposed to normal artificial lighting, if 'after lights out' lighting is now introduced it may seem that the lightmongers are determined that you shall have something switched on throughout the entire twenty-four hours. But for landings particularly - although the time may well come when they are used all over the house - installing fittings which give very dim 'after lights out' lighting is a sound idea. This consists of inconspicuous inset wall fittings mounted about a foot from the floor, containing very low wattage bulbs or fluorescent tubes: the table of burning hours on page 22 will give you an idea of the running costs. Going from a well-lit room you would not be aware that they were giving any light, but for someone who has been asleep or whose eyes have become adapted to the dark they give just enough light to move around, and in a household where teenagers come in at all hours, or small children or elderly people need attention, they avoid the necessity for switching on the normal lighting. To control them all by a single time switch makes them automatic and means that you can forget about them entirely.

Downstairs lavatory

If there is a downstairs lavatory with a wash-hand basin, a fluorescent fitting using a 40 watt 2′ tube in one of the De luxe colours over the mirror will probably be all that is needed. However, to ensure that everything is kept spotlessly clean a ceiling fitting with a 100 watt bulb is a useful addition. If it is a case of choosing between the two, go for the fluorescent mirror lighting since the room will be used by visitors, many of whom will be just as concerned with seeing that they are looking right as with making use of the other facilities. If the filament lighting is switched from the door and the fluorescent pull-cord switched it will ensure that the latter is only used when necessary. For a wc alone, wherever it is, a good strong light from the ceiling is the best proposition and again a 100 watt bulb or two 60 watt bulbs in a shallow enclosed fitting would not be extravagant for the light will only be on for short periods.

Dining room

Now that the dining room has so often ceased to be a room in its own right but is merely a small annex to the sitting room, it has become rather more difficult to light than when it was an enclosed space. The basic rule of a light over the dining table presents no problems, but achieving surrounding lighting which is not going to draw in the walls and look rather ostentatious in a very small annex is difficult. Over the table – and this is one of those places where the plan or model is so useful - *really* over the centre of the table, you can hardly do better than a rise-and-fall fitting that can be lowered so that the bottom rim is below eye level during meals. This could either take the form of an open-topped shade with a 150 watt bulb allowing some of the light to reach the ceiling or a 100 watt reflector bulb in a closed topped shade giving a more positive and directional effect. The concentration of light on the table is flattering to china and glass but is the very thing that makes some form of background lighting necessary.

In a separate dining room, wall brackets teamed with lighted curtains offer as good a solution as any, but you can take your pick from the suggestions in the next chapter. For a small room or annex a pelmet-lit receding design or photomural would work wonders, as would a mirrored wall facing a window outside which there is garden lighting and beside which there are wall brackets. Whether you should provide deliberate lighting for a service hatch to the kitchen is debatable. It is nice to think of each new delicious dish appearing like a prima ballerina on a lighted stage, but not so glamorous as the used plates return to the kitchen. Certainly there should be lighting for a serving table or sideboard, for this need never be used for messy clutter and can be made quite a feature with fruit or flowers or the family silver when the dining table and light are not in use.

Kitchen

If your first thought is 'fluorescent, of course,' it's a good one, but not necessarily fluorescent alone, for even with the best of fluorescent installations an additional strong filament light may be necessary. Basically, however, the aim should be lighting which never makes you throw a shadow on what you are doing. The place where your shadow is most troublesome is in the sink, so the light should not be too far behind you when you are washing up. For a kitchen up to, say, 10' long with the sink at one end, you might just get away with a single 4' twin tube fluorescent fitting with a prismatic diffuser. For anything larger, double up and have one fitting over and parallel to the front edge of the sink and the other roughly in the centre of the room at right angles to the first one. For a really first rate job add properly screened fluorescent tubes under all wall cupboards over work surfaces, and if you want to be really 'Ritzy' install a filament reflector bulb as well over the cooker and sink. If you are not in a fluorescent mood, say, when you are getting a late night snack, you will be glad of the filament light, and if you're simmering a sauce and know that if you miss the first sign of a sluggish bubble it will be overcooked, the additional light will help your vigilance. The kitchen

Fig 15 The blue-lined alcove, lit by sunny coloured tubes, gives the effect of a window. Filament bulbs above the gold louvered ceiling panel light the stairs and the plants. *Author's home*

Fig 16 Most effective lighting has been achieved here with no built-in additions. The bulb in the metal pendant is not visible to anyone sitting down and all its light is confined to the table top. *Courtesy: Homes and Gardens*

Fig 17 In this dining room three fluorescent tubes are used, not only for background brightness but also to highlight flowers, provide a working light for the writing desk, and to illuminate book titles. The pendant over the dining table has a white plastic louver to eliminate glare and the floor standard on the right provides a comfortable reading light. *Courtesy: New Homes*

Fig 18 All cupboards have concealed fluorescent tubes below them and glass bottoms. The lower ones also have glass tops to let the light in. This lighting is used during the day to balance the brightness of the facing north wall. *Author's home*

provides a perfect example of the sort of situation where pull-cord switches can cut installation and running costs and give you the convenience of a switch always within easy reach. In a scheme with under-cupboard lighting, reflector bulbs and fluorescent fittings, all that need be switched from the door – if there are two doors, two-way switched from each – is the fluorescent fitting in the centre of the room. All the rest of the lighting can have individual pull-cord switches and just be used when required.

There is a tendency to think of general kitchen lighting in terms of 4′ and 5′ tubes, but the 6′ and 8′ should not be forgotten. It might be that with the joists running in the right direction you could use a twin tube 8′ fitting recessed between them or, if the width between the joists is insufficient, create your own recess and, using bare tube batten fittings, fill in the bottom of the recess with reeded plastic or small cell louvers. Remember though that recessed lighting will not have such a good sideways spread as a ceiling mounted fitting and that local lighting over work surfaces becomes almost obligatory.

When the kitchen is also provided with an eating area it is a good idea to change the character of the lighting over the table to give a sense of separation from the working part of the room. As in the dining room, a rise-and-fall filament fitting would be suitable. Incidentally, don't forget to give careful consideration to the colour of the tubes you use in the kitchen. One of the high efficacy types such as White, Warm

Fig 20 With a colour scheme like this the wrong filament tube could be disastrous. One of the De luxe types is used here above and parallel to the front edge of the sink and in the wall fitting on the right. *Designers: Vivien Hislop and Peter Rodd*

Opposite
Fig 19 Tailor-made lighting for a tailor-made kitchen suite. The lighting below the wall cupboards is concealed by a pelmet attached to a shelf and not to the cupboards themselves. Two 5′ batten fittings with prismatic diffusers provide a high level of shadowless general lighting.

White or Daylight will do very well. Once you have got used to the slightly distorting effect they have on colours, compared with filament bulbs or real daylight, you will be perfectly happy with them, as long as you had them in mind when you chose the colour scheme. But if your kitchen is gay with all the colours of the rainbow and lots of exciting crocks and cloths, go for one of the De luxe colours, and do see what you think of De luxe Natural. This is a delightful tube, grossly untruthful but in just the right way. It has the amazing property of seeming to make all colours sing – greens look greener, blues bluer, reds redder and yellow yellower. Admittedly if you've been slaving over a hot stove all day you will look rather more flushed, too, but that's a small occasional price to pay for the constant pleasure you will get from everything you see around you.

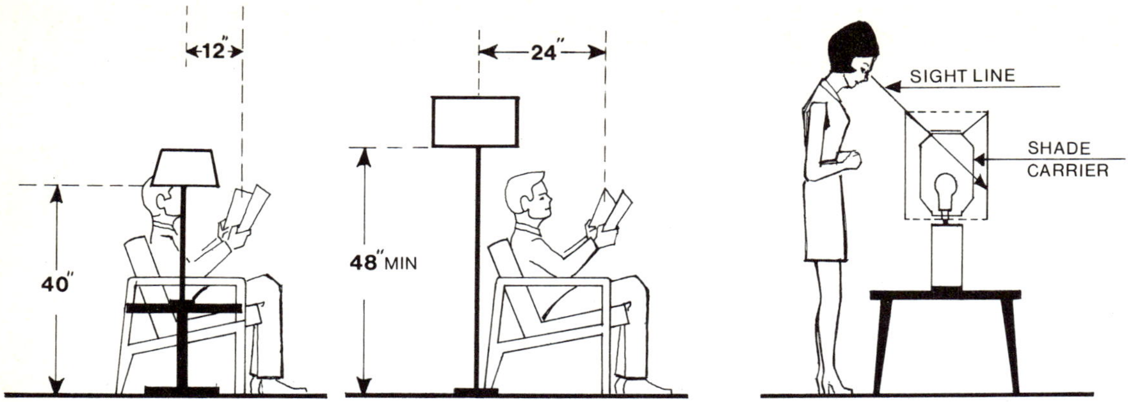

Fig 21 Approximate measurements for floor and table lamps for reading.
Shade carriers are made in many sizes. It is important to choose the right size to suit the base, and to ensure that the bulb is not visible from close quarters.

Utility room

This probably houses the washing machine, spin drier and a double sink. Fluorescent lighting and plenty of it is the answer, and since the room is probably small, one twin tube 5′ fitting will be excellent with white walls, but more strongly directional light from a filament source may be a welcome supplement for ironing since it more positively reveals wrinkles. If you wonder why 5′ tubes are recommended here and 4′ for a small kitchen, it is because in this case you really need to see colours as they are and should therefore use Colour Matching or Northlight tubes. Not only are these among the lower light output types, but being mean with them produces a depressingly cold effect. Be generous and the room will look crisp, clean and hygienic, and if you're brave enough to wash the family handkerchiefs you won't be able to blame the lighting if they are not whiter than white.

Sitting room

Sitting room, living room, lounge, drawing room, call it what you will, this is a versatile room. It must cater for the family and for entertaining and it is hard to draw a line between the basic essentials and sheer fun. In other rooms the lighting is mainly required to do a specific job, to light the sink, the stairs, your book in bed, but in the sitting room appearance matters enormously and the appearance of everything in the room depends entirely on the light by which it is seen. So as well as providing enough light for taxing seeing jobs like reading and sewing, the lighting must be arranged to make the best of everything, to brighten this and subdue that, to provide sparkle or drama or whatever quality is required to make a friendly room. Whether deliberately so or not, lighting is in fact an integral part of the decoration and can enhance or ruin it. This should be an interesting room too, and nothing can provide the

Fig 22 A typical fluorescent batten fitting. It comes in many sizes and forms the core of nearly all concealed fluorescent lighting. It should never be used where it will be visible without some type of attachment to shield the tube.

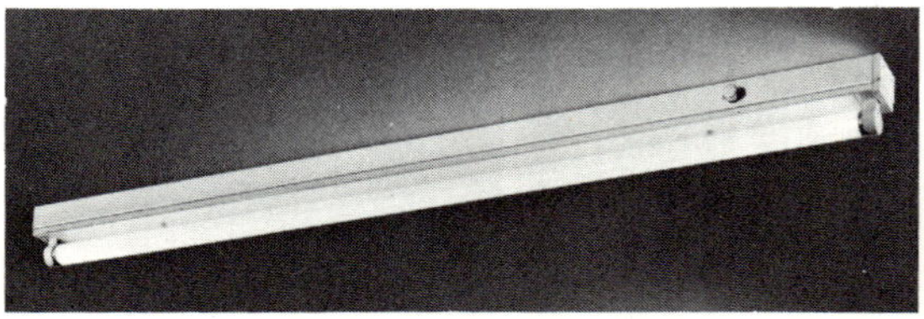

Fig 23 An unusual application of 4′ fluorescent batten fittings each with an additional two straw-textured diffusers to prevent the tubes being seen from the sides. In the intended horizontal position and at the correct mounting height, only the front of the tube would need covering.

necessary variety and emphasis as well as lighting can.

Where ceiling lights are used, remember that they may well be the most conspicuous things in the room and therefore need to be chosen with even more than usual care. But no room could ever look its best when lit by a flood of light from a single fitting overhead.

Other things are needed. While table and floor lamps should be thought about for the appearance of their shades when alight and the effect the pools of light they create have on the overall decorative effect, they are without doubt the best means of providing local lighting at all points where things like reading, writing and sewing are likely to be done. The wattage and number of bulbs will depend on the type of shade but there should never be less than 100 watts in a table lamp and 150 watts in a floor standard with white lined shades on sufficiently tall bases or standards to let the light spread. Directional adjustable fittings with metal shades usually require lower wattages and the manufacturers' recommendations should always be followed.

Fig 24 The large paper sphere, the only visible fitting in this basement room, is only used at night. During the day, tubes concealed around the walls and above ceiling panels give the effect of daylight. *Author's home*

Pictures or flowers can be highlighted by special fittings giving a restricted spread of light. Concealed fluorescent tubes can flood curtains, bookcases, alcoves, the insides of glass fronted cabinets and so on – and in colour if you like. Such lighted features can be used singly or together to provide almost any effect.

As long as the means of producing them are not obtrusive when they are not in use you can hardly have too many lighting ideas in this room, and a wealth of them need not greatly add to the running costs if you are selective. Be subtle; get into the habit of suiting the lighting to the mood of the moment and you will find it is only on rare highdays and holidays that you will want to show everything off at the same time.

Fig 25 There is no need for the elderly to part with well-loved odds and ends if they have to move to modern surroundings. The recessed fluorescent lighting above the window wall and the wall brackets are so functional that they almost cease to exist; the only conscious lighting focal point is the converted oil lamp.

Bedroom

Some years ago, a lecturer on lighting said, 'And now we come to bedrooms where we can have the greatest fun of all', and since in those days he was a rather young lecturer he blushed furiously at the laughter and applause. Today he would say the same thing but rather more circumspectly, for the bedroom can include lots of lovely nonsense that you might not dare to expose in the more public parts of the house. But here, of course, we are concerned with the practical side of things and this can very briefly be summed up by saying that there should be at least one fitting two-way switched from the door and near the bed for general lighting, individually switched bedhead lighting for each person, good lighting at the dressing table mirror and the long mirror if there is one, and lights controlled by automatic door switches in wardrobes. A floor or table lamp beside an armchair is useful, if for nothing else, for those

Fig 26 The most interesting aspect of this room is the switching. A touch of the switch on the right, which is by the door, turns on the pendants; if it is held for a second or so, all the lights come on. With the switches by the fittings, it is the nearest fitting which lights first. Switching off is the same procedure in reverse, except that the quick touch by the door turns everything out. The white panel beside the dressing table is a low wattage 'after lights out' fitting.

times when you are up but still bedroom-bound after an illness, although in these days of central heating being 'confined to one's room' is more of an excuse for not taking telephone calls than a necessity. Again for your bedroom a plan or model can be of the greatest use, since both number and position of socket outlets will be of lasting importance to you. Two single beds need not be side by side, dressing tables need not block the window, neither is there any need to kill the usefulness of a wall by having a fixed fire or heater.

Ideally all the bedroom lighting should be two- or even three-way switched from the door, a position near each fitting and from near the bed. This is often done in hotels, but in the home, if socket outlets are to be used, it is seldom practical, and one or two fittings that

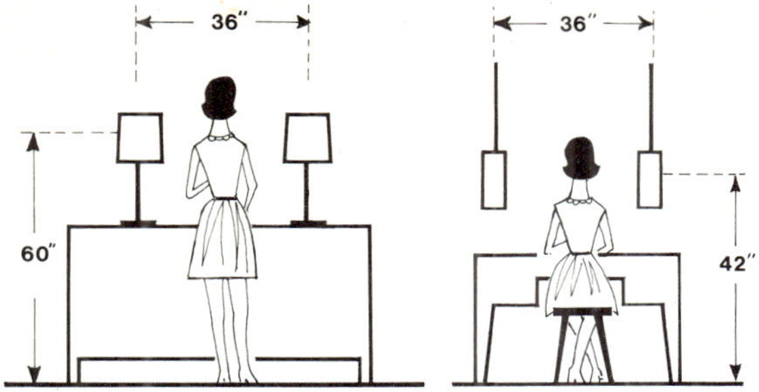

Fig 27 Suitable positions for table lamps and pendants at mirrors when either sitting or standing.

can be two-way switched from both the door and bed will fill the bill with all other lighting switched individually. Elaborating on this, although it may sound a bit far fetched and, of course, is quite a considerable extra expense, for nervous people or those with houses in isolated positions, a two-way switch beside the master bed operating, say, the hall and landing lights and one in each of the downstairs rooms, is a great consolation if things should go bump in the night. It would be a foolhardy burglar who would not make a quick getaway if a sleeping house suddenly woke up.

In a softly feminine room by far the best way to light the face – remember, the face not the mirror – is to have a pair of table lamps on the dressing table which raise the bulbs to cheek height. Use white shades so that you will have a reasonably truthful light for make-up and at least a 60 watt bulb in each. Half-shades that can be turned to expose the bare bulbs can be used as long as the bulbs are not reflected in the mirror with the shades in their normal position and do not make the area around the dressing table too bright in comparison with the rest of the room. A dangling fitting alone above the head is utterly forbidden, but if you can find one that will take a 100 watt bulb and go happily with the table lamps and look intentional it can be a good wrinkle-detector.

In a severely male bedroom a fluorescent fitting with an incorporated electric razor socket can be mounted at the top of the mirror. Time was when one would have stressed the importance of good mirror lighting for a man in order that he could see to shave. In these days of male make-up it is even more necessary, for while it is well-known that many men use some form of bronzing agent, the man who looks as though he is wearing make-up is still 'taboo' and his only safeguard against this is to be able to see his face in every grisly detail. A cold, truthful Colour Matching or Northlight tube is best if his nerves can stand it first thing in the morning.

Since wigs and changing hair colours also seem to be here to stay, it is worth giving them a little thought here. If you want to be sure that your hair outline is correct, try to arrange a dark background opposite the mirror for light coloured hair, and vice-versa. It's a bit of a shock if you've been lacquering away like mad to catch sight of yourself when you're far from your spray and find that you're lumbered with a rigid whisp which you can do nothing about for the rest of the evening.

If you can take care of your face elsewhere, light from shoulder to toe should be the aim at long mirrors and can best be provided by a wall mounted fitting on each side: 3' or 4' filament architectural tubes would do, fluorescent tubes would be better. It is a great advantage to have an additional strong downward light overhead to show up creases. A ceiling inset downlighter is the least obtrusive way of providing this for a ceiling mounted or pendant fitting may not be all that easy to fit into the décor.

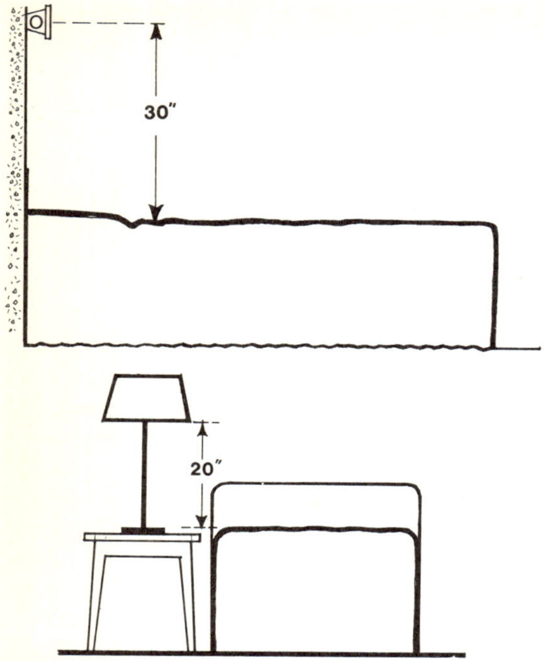

Fig 28 Use 100 watts in bedside reading lamps. A 2' 20 watt fluorescent tube can provide an alternative at the bedhead.

Fig 29 An adjustable twin reflector bracket suitable for mounting centrally on the wall above the head of a double bed. With separate switching, the light can be directed away from the person who wants to sleep.

The occupants of double or twin beds, whether side by side or not, must have separate bedhead fittings, each of which lights the other person as little as possible. Table lamps always look pleasant but the most foolproof method of restricting spilled light is to have a pair of opaque-shaded adjustable directional fittings mounted on the wall between twin beds or over the centre of a double one, about 30" above mattress level. Never clip anything to the headboard. If this is tall and ornate it will look ghastly unless the lighting has been specially designed, and if it is of the normal height and you are sitting up you will probably find it rather hot on the back of your neck and it will cast a shadow on your book. The mistake most frequently made with bedhead lighting is to mount it too low, and if normal wall brackets are being used, to place them centrally overhead. They should be mounted above the edges of beds, and for all types of lighting the 30" above the mattress – not pillow – rule should be followed, to allow a correct light distribution (see fig 28). Aim at a 100 watt bulb for each person, although a very directional type of fitting may require less.

Unless the position of the bed is quite definitely known, most forms of bedhead lighting are best plug-connected and finally fixed when the furniture has 'settled'. A small amount of

Fig 30 A successful combination of concealed and visible fittings provides a wide variety of lighting effects. Each is separately switched, and the mood can be endlessly changed. *Designer: Charles Kenrick*

visible flex is much better than a firmly fixed fitting which looks wrong because the bed has had to be moved even if only a few inches. Always have a twin socket outlet for each person, not just for lighting but for electric blankets and tea-making appliances. Where the exact position of the bed is known, this is one of the instances where switched sockets mounted well up the wall, possibly just above the tops of bedside tables, will be a great convenience.

In rooms with a loft above, it is quite easy to have ceiling points installed wherever you require them, and, if only you can find an electrician with clean hands, they can be put in after you have finished the room. Two pendants 3′ apart could light the dressing table mirror and you could use pendants over the bedside tables instead of wall brackets or standard lamps. If you use pendants, choose an unbreakable light-weight shade material rather than glass, for they are bound to get knocked or start to swing if a window is left open.

Children's rooms

Fig 31 Orange card shades around internal diffusing glasses are used on the pendants and on two table lamps elsewhere in this room. The bedside fitting has a rise-and-fall appliance in the ceiling rose. To give a greater sense of space there is no door and glass is used for the table top and below the plants, offering no obstruction to light or vision. *Author's home*

One sees many very pretty illustrations of frilly night nurseries with whimsy wallpaper, but to install lighting solely to fit this type of décor is a big mistake. Babies very soon become people and the basic lighting installation should allow for the change. You will, naturally, have to get different fittings as time goes by, first something tough and unbreakable to cope with the hooligan stage and then another retread when the room becomes a self-contained bed-sitting room, but there should be no need to make wiring changes. Two fixed fittings are all that are needed to begin with, a wall bracket in the bedhead position (sink the wiring in the wall in this case so that it can't be played with) and one ceiling fitting, placed possibly to the left of the window for the homework/hobbies stage. In front of the window is a good place for a useful table and a rise-and-fall fitting will do the job of a table lamp without the risk of its being knocked over. Be very liberal though with socket outlets, for this room is going to

Fig 32 The fluorescent-lit ceiling panel, patterned and multi-coloured, was an addition to this bathroom. The original high ceiling has made it possible to drop it sufficiently for two 6′ tubes mounted about 18″ apart to give an even light distribution.

have to cater for every kind of electrical appliance, irons, hairdriers, tape-recorders, record players, TV, and so on, and it is safer for them to be plugged in correctly than to spring from a growing tree of multiple adaptors, for however restricted the wiring and whatever parents may say, used they will be. Incidentally, if this room could be soundproofed from the very beginning it might be a great blessing all round. Thinking, though, of the early days, it would be wise to make quite sure that these outlets are of the shuttered type so that small fingers cannot explore them. One of the low level 'after lights-out' fittings will also be useful for a number of years.

Lofts

With children still in mind, loft space can often be turned into an admirable play or hobbies room. To say how it should be lit would mean running through the entire gamut of conversion treatments. As a start, temporarily install perhaps three double socket outlets each with plenty of spare cable; these should satisfy most needs, and when the time comes to board the floor and line the roof at least you won't have the added expense of additional wiring. Incidentally, of course, all lofts should be provided with at least one light switched from below.

Bathroom

This is the one room in the house where you ought to be able to see the very worst of the whole of yourself in the most unflattering light. Its décor should be largely white and a twin 4′ fluorescent fitting with Colour Matching or Northlight tubes on the ceiling and another tube of the same colour at either side of the mirror are really the least you should have. Supplement these with a ceiling inset strongly directional filament light overhead for merciless detection of anything the fluorescent lighting might flatten out. Do try to make it a longish mirror. If you place it over the basin this will be

Fig 33 These 4′ fluorescent batten fittings have reeded fronts and white opal sides; they do not look over-bright since they are mounted on a white wall. Operation is by a pull cord switch on the end of which is a brightly coloured wooden ball.

Opposite
Fig 34 Bare Colour Matching fluorescent tubes give ultra-truthful lighting at the mirror. A second inset ceiling fitting is used over the bath and another Colour Matching tube is concealed by a pelmet above the basin. *Author's home*

difficult but it is seldom necessary to have it there, and if increasing flabbiness and fat are to be noticed before they go too far it really is necessary to see more of your naked self than just your head and shoulders. As long as they are correctly installed, built-in wall switches are permissible, but if in doubt use the pull-cord switches that are normally recommended.

Of course, if you don't want to face facts in your bathroom then you can flatter yourself to your heart's content. Use one of the 'kind' tubes behind diffusing panels in walls and ceilings, see yourself against a background of cerulean blue, and glowing health and youth will seem to last for ever! But surely it's better to know the worst in private than let other people discover it for you in the cold clear light of day.

5 Miscellany of ideas

All homes can be lit successfully with visible fittings, if you shop around carefully and know what you are looking for; but whatever their shape, size, period or style of decoration, they can all be more fun to live in with the addition of a little lighting sleight of hand. Some of it is complicated, but most of it is unbelievably simple, and there is nothing in the following miscellany that is not a perfectly practical proposition and, furthermore, probably very little that you have not had the chance of seeing for yourself. These are ideas that are in constant use in shops, restaurants, cinemas, theatres; on the stage, on the screen, in television studios; and while in these places they may be applied with rather more bravura than would be suitable in most private homes, anyone who takes the trouble to look can see for themselves what can be done.

Colour

The colour-rendering properties of fluorescent tubes have already been mentioned, but now a word on room colours and their effect on lighting. The lighter a colour is, the more light it will reflect, that is obvious; but many people do not realize that coloured materials appear to be a certain colour because they can reflect that particular colour of light but absorb other colours of light. The overall colour of light in a room is made up of what comes out of the bulbs and tubes or through the windows, plus what is reflected and re-reflected by the room surfaces. Suppose you have a room where green is used extensively. Both daylight and any kind of artificial light contain plenty of green which the walls will reflect, so the dominating colour of light will tend to be more or less greenish. While this may not be worrying during the day it can be most upsetting at night if much of the artificial lighting depends on reflection from the surroundings. For example, a kitchen decorated predominantly in green or yellow and lit by high efficacy tubes, which are deficient in red, would make everything and everyone look decidedly seasick. However as long as you are aware of what reflection does to the colour of the light it can be turned to great advantage. Dull rooms can be made to look sunny; over-bright ones, cool; cold ones, warm and so on. Fashion in room colours changes enormously. At the time of writing matching patterns of wallpapers and fabrics are popular and so are all-over dark colours for walls, ceiling, floor and woodwork, whereas a few years ago a logical and sensible use was being made of different colours for different walls. This still seems a sound idea and if you have made a three-dimensional model of a room it is interesting to experiment. There is a wide range of coloured papers in most artists' supply shops, or you can create your own with poster paints on fairly stiff paper. This is better than painting the model itself since quick changes then become difficult.

Fig 35 Matching dark green walls, ceiling and floor provide little or no reflection, so basic lighting is directed solely at the brightly coloured furniture. Additional lighting, for use as the mood demands, is provided by a back-lit brilliantly coloured wall panel, an unusual pillar floor lamp and two low-voltage spots which create pools of light beneath the translucent table top.

Never underestimate the value of white, but beware of floodlighting a white wall from a light source only a short distance in front of it; the light will arrive at a fairly glancing angle, showing up shadows and emphasizing the least irregularity in the surface. If you want to use concealed directional lighting and a light colour on a faulty wall, cover the wall either with some strongly textured material or a boldly patterned one. White is also wonderful for framing things: you cannot go wrong with it for door frames and skirtings. Use it too for its ability to distract. For example, a dark coloured wall framed in white and with white objects placed in front of it – white opaque lamp shades on white bases, a white vase or chair or frame for a mirror or picture – and no direct light falling on it, will seem to recede out of existence. Light should never be deliberately directed towards dark coloured surfaces, for their main value is as a receding background. A light-coloured ceiling with them is perfectly permissible and light can spill upwards to this, but the main object should be to focus lighting on white or brightly coloured objects in the foreground. If the ceiling is dark coloured too, then direct all light downwards. This technique is most successful for town rooms. Where sun and light and gardens form part of the daytime scene it is inclined to look over sophisticated.

Really strong coloured light, such as you get from the green, blue, gold, pink and red fluorescent tubes, can be most exciting for occasional use. It is best to use it for concealed background lighting of white walls, but it can also create spectacular colour changes on coloured walls or curtains; though care is needed if the overall result is not to be rather muddy. As with all lighting for effect, experiment on a small scale before you commit yourself.

The coloured dichroic bulbs are splendid if you require coloured shadows, not only of solid objects but also, say, in the folds of drapes. They can produce wonderful effects. Remember though that the primary colours of light are red, green and blue, not red, yellow and blue as in pigments, and that with coloured light you need to overlap green and red to produce yellow, green and blue for turquoise, and red and blue for magenta. Of course if you overlap all three in equal amounts you will get white, but don't let this deter you, for you obtain coloured shadows by using a bulb of one colour to light surfaces shadowed from light of another colour. Supposing, for example, you throw a red light on to a white figurine from the right and a green light from the left. You have then created two shadows. The one caused by the red light will have no red light in it and so be lit by unmixed green, and that caused by the green light will be lit by unmixed red, but all around both green and red will overlap. The end result will be a yellow figurine with red and green indentations, the whole seen against a yellow background with green and red shadows, which can be far more attractive than it sounds. Bulbs give the best shadows; generally fluorescent tubes give too diffuse a light for effective shadow-making.

Coloured PAR bulbs or dichroics in inset ceiling fittings can colour different parts of the room. For example, red could be used for the area in front of the fireplace, but unless the sources of the coloured light are intended to be seen as part of the décor, it is best to use 'black hole' fittings; with these the coloured light

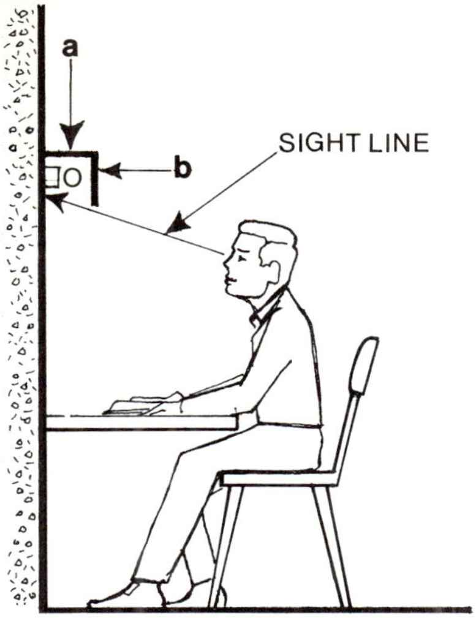

Fig 36 For any type of pelmet lighting, work out the sight lines from normal viewing positions – usually sitting, but for an open-top pelmet, standing as well. **a** and **b** should rarely be less than 6" but in a case like this the sight line will also help to give the correct height from the floor. Assume that the sight line is 20° above the horizontal when sitting and 20° below when standing.

below is all that is noticed. Where this treatment is used keep it away from places where people will be sitting for it may have unkind effects on clothes and make-up.

Concealment

Pelmet, valance, facia-board, baffle – all these terms are used to describe the most universal method of concealing bulbs and tubes; to avoid confusion they will be referred to here as pelmets. Basically a pelmet is a strip of anything which is wide enough to prevent the light source from being seen and far enough away from the surface to be lit to allow the light to spread either upwards, downwards or in both directions at once. It can be mounted horizontally or vertically and is foolproof as long as the sight lines have been worked out in advance. Mistakes can easily be made but equally easily avoided either by experiments or drawings. When doing the latter, accurate scale and the use of a straight edge are important, for light travels in straight lines and even half an inch too little on a pelmet, revealing no more than a sliver of one side of a tube, may be quite enough to spoil the entire effect. Where tubes under a pelmet run across a doorway, French windows for example, then some form of cross baffle should be used. Clip-on plastic ones are obtainable but they can also be made up in a thin material like hardboard and painted – a colour to harmonize with the curtains or wall will make the louvers less noticeable and will not substantially alter the colour of the light. Another method, where the pelmet is being used to light a wall or curtains where the heading needs to be seen, is to fill in the bottom of the pelmet with louvering. In this case plastic 'metallized' louvers are good for their low-brightness. They can be obtained in silver, gold and bronze, both shiny and matt. An alternative method, if the joists run in the right direction, is to stop the ceiling short of a wall but allow it to project sufficiently into the gap to hide the tubes from anyone looking upwards. Where more than one tube is needed to complete the run under any sort of pelmet, the ends must overlap an inch or so. Just butting them together will create a noticeable shadowy patch and while this may not matter so much for curtains, it is disastrous for walls and ceilings. This is one of the few cases where it may be necessary to use tubes on their own with the control gear concealed elsewhere, since leaving room to overlap batten fittings may mean that the pelmet has to stick out uncomfortably far.

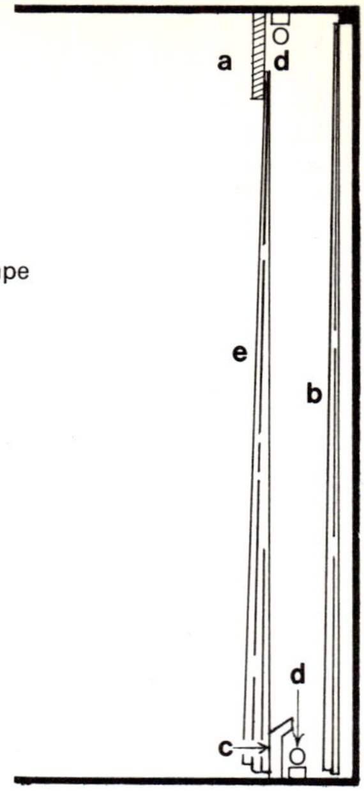

Fig 37 **a** Pelmet
b White iridescent drape
c Floor pelmet with angle return
d Colour-wrapped reflectorized tube
e Fine white net

Another point to beware of with pelmet lighting is reflection in glossy surfaces. Never use a pelmet on any glossy, polished or mirrored surface for it will be virtually impossible to avoid a reflection of the tube and this will cancel out all the trouble you have taken to conceal it. For the same reason, any form of indirect lighting with a glossy or even slightly glossy ceiling is a non-starter. If you must use pelmet lighting for a wall mirror, mount the pelmet sufficiently far above the mirror so that the light is reflected from a matt light-coloured surface – white for preference because the light will be tinted by the colour of the wall.

Pelmets can be perforated and translucent, and, while to create a long unbroken line it is best to have them specially made, there is a wide variety on the market which can be butted together, although with these, of course, there will be the 'dead' patch where they meet. For all pelmets which light in one direction only, it is worth getting hold of 'reflectorized' tubes. These have an internal reflector which cannot become dirty and which directs practically all the light through one side of the tube. An ordinary tube in a white-lined pelmet would be almost as effective at first, but would become less and less so as the white lining collected dust. Unless you look twice it may not at first be apparent which is the 'light' side, so beware of putting it in facing the wrong way. If you are using one of these tubes, or any tube for that matter, for lighting a white wall, you can get wonderful effects by wrapping sections of the tube in coloured 'gelatines', the sort used for theatre lighting. Use a single thickness and, overlapping at the top, hold it in position with thin wire; Sellotape soon comes unstuck from the warmth of the tube. (Make a sandwich of fine white net on the room side, a colour-wrapped tube and white iridescent drapes against the wall and you'll be delighted at the result.)

While a pelmet which is open at the top and bottom will let out light in two directions, there are times when it would be nice to ring the changes, and this can be done by having a horizontal divider with separately switched tubes above and below it, or the pelmet board can have centrally placed spindles at the ends held by projecting brackets. In this way a single tube can light upwards or downwards or in both directions at once.

Panels

Where either bulbs or tubes are used behind translucent glass or plastic panels, the pelmet system is again useful around the edges. If, however, the panel is too large for the light reflected from the internal white background to

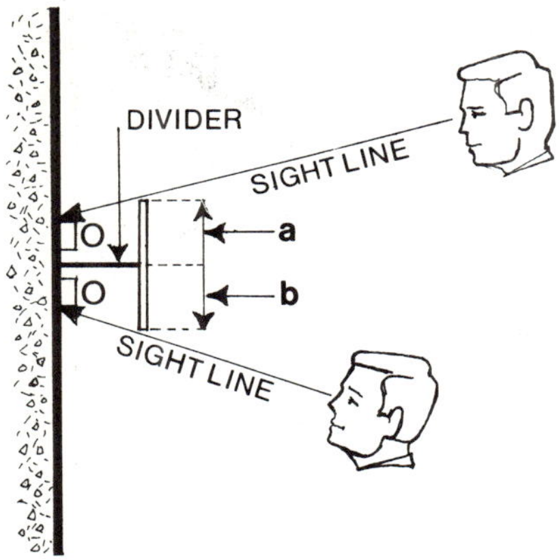

Fig 38 Separately switched tubes in a divided pelmet. Sitting and standing sight lines should govern depths of **a** and **b**. The smaller **a** is, the farther light will spread across the ceiling.

reach the centre, the light sources will have to be placed behind it. This, if you are not careful, can give a spotty or streaky effect. For bulbs the spacing should not exceed 1 $^{1}/_{2}$ to 1. Thus if you have only enough depth for placing the bulb centres 6″ behind the panel, they should not be farther than 9″ apart both sideways and up and down – for a large panel with little depth this is a hot business needing a great deal of ventilation and best forgotten. The spacing for tubes is the same and although they will not produce nearly so much heat, plenty of ventilation is still necessary for they can become hot after a time.

If a lighted panel in a wall, floor or ceiling is something you have set your heart on and you cannot avoid spottiness or streakiness, try to superimpose some design that will look intentional. In a wall panel, for instance, using glass or sheets of plastic, you could have leaves, pressed flowers or coloured tissue paper or acrylic shapes very lightly stuck to the front of the obscured sheet and then covered by a clear sheet. The same idea using two clear sheets is fun where the tubes are out of sight around the edges. Once again, experiment but don't rely on the pressure of the sandwich to hold things in position. It's heartbreaking to make a beautiful horizontal arrangement only to see it all gradually slide to the bottom when in a vertical position. This will not apply in ceiling panels or if you use a luminous panel for a table top; small box-like luminous panels are easy to have made up for lighted bases for flowers, bottles or glass. Translucent material with laminated 'patterns' is available if you feel that making your own is beyond you. Lighted panels are often useful as a background for shelves on which to show things in silhouette.

Another way in which a luminous panel can be used is as a shelf. Here, however, the treatment is different, for even if it is below eye level it will be seen sideways; thus uniform brightness is not important and a single tube can be concealed by the opaque front edge. For a bright shelf the underside should be filled in and painted white inside so that the whole thing forms a shallow box with a translucent top. If the shelf is wider than about 12″, the bottom can be sloped upwards towards the back. This will increase the brightness at the back by bringing the reflecting surface closer to the luminous top. If the luminous panel forms the top of a low built-in cupboard and the reflecting surface is translucent, the system can double as cupboard lighting. Should no reflecting surface be used it is essential that the inside of the cupboard is painted white for the top panel to appear to be lit.

Keep lighting in mind, too, for those shelves above heat storage units and fixed radiators, the purpose of which is usually to prevent convection currents marking walls. Making the shelf into a luminous base for ornaments will distract attention from what is below if, as so often happens, it is not a thing of beauty. When radiators are below windows the sill can be extended and made luminous for upward curtain lighting and, if it is of good strong glass, provide a base for plants. Because of the extra heat from a radiator it is probably better to use architectural filament tubes with heat-resistant wiring, since fluorescent batten fittings do not like too much heat and finding a home for separate control gear makes everything rather complicated. For all lighted shelves which have to carry any weight use a thin diffusing material – flashed opal glass or white plastic – under the required thickness of clear glass. Obtaining a diffuser sufficiently thick to carry weight, let light through and not scratch easily is well nigh impossible.

To create the feeling of a shelf floating in space below eye level – remember, eye level when seated – leave a gap of at least 3" between it and the wall and sandwich the tubes – fluorescent or filament – between the back edges. They should be placed as near the back as possible without being visible so that the light can spread. If the supports are slender and set into the wall their shadows will not be noticed. This is a good idea for a fireplace wall.

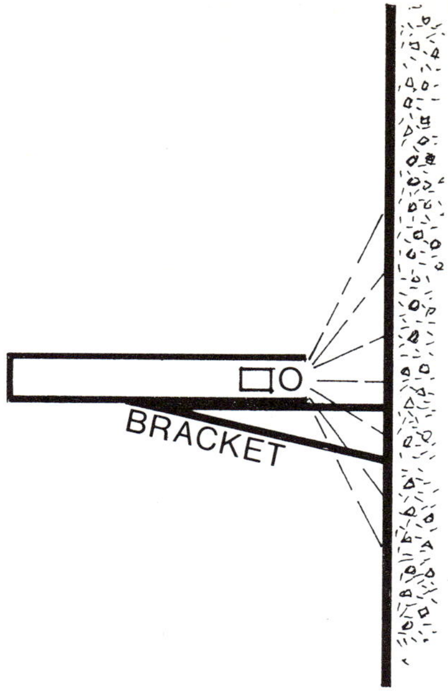

Fig 39 Section of a 'floating' shelf. The fluorescent tube at the back is about 3" away from the wall and throws virtually no shadow of the slim supporting brackets.

Enlargement

In the section on kitchens mention has been made of undercupboard lighting for work surfaces, but there is another aspect of this which can be applied in other rooms, particularly small ones. If you have a cupboard at about eye-level, it tends to reduce the size of the room by the distance that it projects. Concealed lighting above, below or in any gap beside such a cupboard draws the eye subconsciously to the wall surface behind it; consequently, instead of getting the mental impression that the room stops at the cupboard door, the eye is carried back to the real size of the room. Light inside a cupboard will also do this to a certain extent and if the doors are solid and the contents pleasant to look at, it may be a virtue at times to leave the doors open, just to give this extra

Fig 40 Three interesting ideas. The hollow supports to the shelves on the left contain slim fluorescent batten fittings; the telephone table has tubes behind it; and the space behind the right-hand shelves – fixed some 3" away from the back wall – is filled with white translucent opal plastic in front of two fluorescent tubes which backlight the panel as well as the wall above and below.

stretch of vision. The effect can be made more interesting still if the inside of the doors is given a completely different treatment from the outside – colour changes, the use of posters, exciting wallpapers. In rooms with walls of floor to ceiling cupboards a greater sense of space can be created by making the doors into luminous panels using either translucent material or slatted louvers. They both give the impression that there is something beyond.

Nothing can beat mirrors as room enlargers, but always plan reflections carefully. A lighting fitting which looks fine in itself may give quite the wrong effect when reflected. This can apply to wall brackets mounted on mirror-faced walls. Not only do you risk spoiling the shape of something that is designed to be seen from the front and once only, but it can also break the continuity of the reflection of the room. One of the most effective ways of increasing the apparent size of a room with mirror is to use it around the outer edges of a wall in strips of about 18 inches to 2′ in width and have plenty of light on the facing wall. In this way the game isn't given away immediately by people seeing themselves reflected, but they have the feeling that the room goes on behind a screen on the mirror-edged wall. This is a good idea for a small hall in a small house. Someone entering it for the first time gets the impression that it is twice the size and although they may very quickly realize their mistake, the first impression stays. Also, where people are not going to see themselves reflected, strips of mirror opposite each other can mislead the eye. One on the door wall with a spotlight above it lighting flowers, or a cluster of fittings hanging in front of it, will first be noticed by its reflection in the one on the facing wall and before you know what is really happening you will have the feeling of a room well over twice as long as it is.

Carrying the eye to the room's extremities can also be achieved by lighting behind furniture and, supposing you want to silhouette objects on shelves, sliding light behind the shelves. This is another use for the ubiquitous pelmet, but always remember that it can't produce miracles unless you allow enough space for the light to get out; and where furniture or shelves are being back-lit space is again needed for the light to get behind them.

But with large windows on the ground floor one need not be limited by the room's extremities. During daylight we do not consider the area outside the windows as being related to the interior; but if it is lit at night this area seems to become an extension of the interior. Filament bulb fittings can be used here, and certainly should be if the light has to travel more than a short distance; otherwise, a simple and unobtrusive method is to mount a fluorescent tube outside immediately above the window. Where there is some projection above the window or where a simple pelmet-like cover to give protection from the weather can be built, an ordinary batten fitting may be used, but a tube with waterproof holders, and with control gear mounted inside the house, is probably as good a method as any, although a small projecting pelmet above this, too, is a worthwhile addition to reflect the light downward and to keep it out of other people's bedroom windows. There are special exterior fluorescent fittings for garden lighting but they are scarcely things of beauty for clamping on to the side of the house, and the pelmet idea will provide the pool of light close to the window which is the object here. Where there is a patio or porch roof, straightforward single tube fittings with diffusers, again on the wall above the window, are best, for they then look a deliberate part of the scene when sitting out of doors. Visible filament bulb hanging fittings are not so satisfactory for giving the room extension effect since they punctuate the 'flow' feeling when seen from inside.

At one time room dividers were the 'in' thing. There are still many of them around but, unless they serve some useful purpose, curtains or sliding partitions lit by inset ceiling fittings are just as effective. Open and with the lights off there is no break between rooms, and

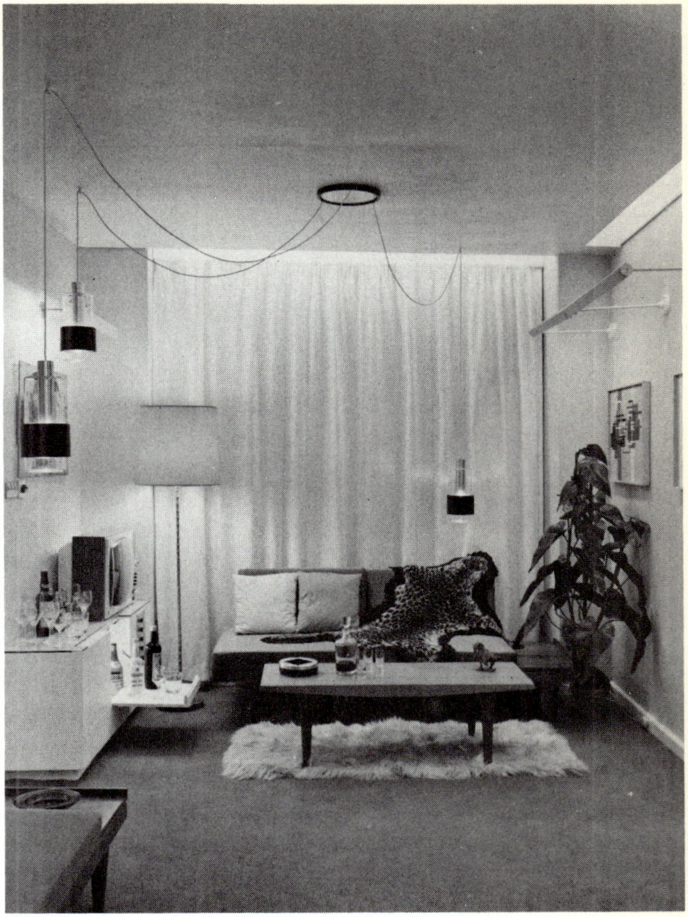

Fig 41 Lighting of the curtains and right wall is from tubes above inset diffusing panels. As an alternative for the latter there are two fluorescent picture-lighting fittings mounted end to end. Not only are the curtains white – for maximum lighting effect – but the lighting itself is far enough away at the top for the light to spread downwards. Long flexes make it possible to loop pendants from a standard ceiling plate to strategic points. The drinks cupboard is internally lit.

closed with lights on a very emphatic one. If there is a divider, the lighting depends entirely on its design. Narrow pelmets for concealing slim tubes can be attached to shelves, ceiling inset fittings can highlight plants, there can be horizontal or vertical luminous panels. Table lamps look well, too, if there is room for them, but if they are to be used it is a good thing to have socket outlets built in to the divider itself for these and any other electrical things it may carry, since flexes trailing to it from nearby wall outlets can give it a tatty make-shift appearance.

Looping

Where any part of the house has insufficient ceiling outlets for pendants in the best positions, the cost of new basic wiring can often be saved by 'looping' a number of fittings from a single ceiling point. This simply means what it says, taking the flex in a graceful loop to an insulated hook over the point where the fitting is required. As was suggested for the hall, a special spreader can be made up but they can be bought ready made, or it may be possible to adapt a branched pendant simply by lengthening the flexes. Professional advice should be taken on electrical loading of the wiring to a ceiling point intended for perhaps 200 watts at the most, as it may become overloaded if it is expected to run too many bulbs. In the right rooms looping looks far more intentional than running wiring on the ceiling and trying to pretend that it doesn't exist. Another method is to take a flex up the wall from a socket outlet and start looping where it hits the ceiling. Sometimes this can be linked with ceiling mounted or inset curtain rail and a number of loops attached to runners so that the fitting can be moved across the room.

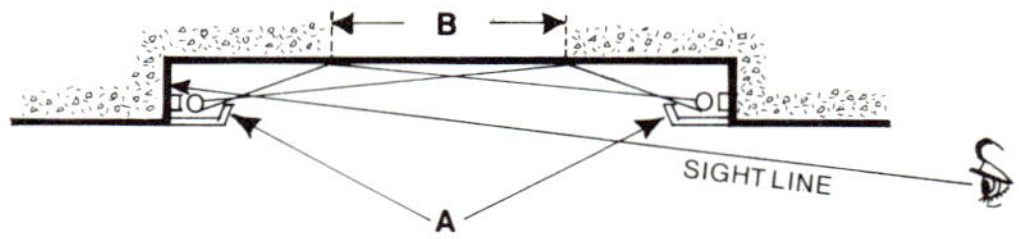

Fig 42 Vertical pelmets at the front of alcoves need returns A to prevent the tubes being seen from the sides. The returns should be as narrow as possible so that light from the tubes can spread to overlap at B and give a uniform background brightness.

Alcoves

Very few new houses have ornamental alcoves or niches, but they can do much to liven up conversions of old ones. Where a chimney is no longer required, it may be possible to scoop out an alcove above the mantle-shelf, or the existing recesses beside a chimney breast can be levelled off with blockboard into which alcoves can be set. Use pelmet lighting around the edges and if the top is to be curved, tailor it to take curved filament architectural tubes. No type of domestic tube can be bent to your requirements, so let the curved tubes that exist dictate your design. Unwanted cupboards and doorways are also suitable places for alcoves which make good settings for flower-pieces and ornaments. If they contain shelves it is best for these to be made of glass so that they don't interrupt the flow of light. Another entertaining treatment is to have lighted bases as well as lighting around the edges. The two systems can be separately switched, the sides for fluorescent tubes and the base for coloured filament bulbs. Using a simple thermal flasher, two colours of bulb could be concealed in the base and made to come on alternately – not such a garish idea as it may at first sound. Whilst on the subject of automatically changing lighting, it is worth mentioning that, as well as apparatus that will constantly merge one colour of light with another, there are now on the market a number of devices that can be linked electronically with radios and record players to create rhythmic lighting patterns on a special screen or even make different fittings in a room dim and brighten to the pitch, volume and rhythm of the music. The screen can be as small as about one foot square and could be set into a wall, but for fun and games with fittings – which must contain filament bulbs and not fluorescent tubes – you need a large space. Naturally, this isn't everyone's cup of tea but imagine it applied to underwater lighting, fountains and floodlighting for a pool-side party and it certainly has possibilities. Incidentally, if you're in this income bracket, don't forget that there are now exterior fan heaters designed to take care of this sort of thing if there should be a drop in the temperature.

Fig 43 Vertical pelmets light the sideboard alcove and there is a 'floating' canopy above the dining table. Where indirect lighting is required, a suspended structure of this nature can often be more effective than cornice lighting round the walls. There is still the opportunity for suspending fittings if the room demands them.

High ceilings

In the right sort of room a high ceiling presents no lighting problems. A spectacular pendant can be used as a space-ornament with surrounding concealed indirect lighting from cornices or reflector bulbs in floor standards; gilded or silvered finishes to mouldings on a deep colour will look marvellous with no special lighting but just the unplanned spill from portables. In a large lofty room with modern décor a good gimmick for lighting the ceiling indirectly and evenly where light from the sides will not reach the centre is to suspend V-shaped beams on rods with the electrical wiring entering from the walls, or at a single suspension point if they do not touch the walls. If the bottom of the V is no wider than necessary to house a slim batten fitting and the splay does not exceed 15° from the vertical it will be hardly noticed and the effect will be that of a very slender structure since the eye will be deceived into seeing each beam as no wider than the bottom surface. The splay is necessary

Fig 44 In this conversion the centre of the ceiling has been lowered to give depth at each side for a leaf-like luminous ceiling and the alcoves beside the fireplace have been backed with large lighted panels in front of which are luminous shelves. Spotlights above white louvers highlight the abstract shape on the blockboard-covered chimney breast.

to allow the light to spread and working out the lines of cut-off on a drawing will show how much it should be as well as the depth of the V; as with lighting behind luminous panels, the farther the beams can be dropped from the ceiling the farther apart they can be spaced.

But if rooms in large old houses have been subdivided to make flats, either lowering the ceiling or creating the feeling of a lowered ceiling is often necessary. A completely false ceiling provides opportunities for insetting lighted panels and downlighters, and concealing lighting around the edges, but should be given careful thought before it is put in, for in the early stages it is easy enough to vary the position of supporting battens so that they do not prevent the insertion of lighting exactly where it is required. Leave the lighting until too late and ten to one where you would like to have an inset fitting you will find a batten. Also there must be provision for fixing the lighting and it may be much cheaper in the long run to use blockboard rather than hardboard, plasterboard or plywood. Remember, too, that if the lighting is being concealed around the edges you need an unbroken line and wall fixings for the ceilings will have to be arranged accordingly.

Should an all-over generous amount of lighting be required in such places as kitchens, bathrooms, games rooms etc., an entirely luminous ceiling may be just the thing. They take many interesting forms – diffusing panels, wedge-shaped translucent tiles of various orientations, double skin sound-absorbing squares, small-cell louvers in both white and metallized plastic, and vertically hanging metal leaves. With the last two, as with any form of grid, the full air space is still, as it were, in circulation, which may be a disadvantage economically because a sealed lowered ceiling, at any rate theoretically, reduces the air space to be heated. There are also types like very large diffusing fittings which can be hung on chains but with these, indeed with all of them, tube spacing above is important and manufacturers' design departments should be consulted since, as with luminous panels, the wrong spacing can ruin the effect. A luminous ceiling is quite a good thought for a high narrow hall where it could not only reduce height but give sufficient light to avoid a feeling of gloom when coming in from a bright day. At night a separately switched planned pattern of filament bulbs above the ceiling could be substituted for fluorescent tubes. When lowering a hall ceiling always take care not to restrict head room at the foot of the stairs.

Cheaper than a complete false ceiling is to paint the existing ceiling black and suspend a white-painted open grid below it. Sections of the grid can be filled to take downlighters, pendants or flush fittings, or pendants can hang through it from the ceiling above. If the grid is to give the impression of a complete lowered ceiling it is essential that no light is allowed to spill above it. Another way of using the grid is to have black or very dark-coloured socket outlets attached to the ceiling and, using dark-coloured plugs and flex, run this to unfixed solid panels which are simply resting on the grid and use these for pendants or inset fittings. In this way you can alter the lighting pattern as you like. Cross baffles may also provide a solution but they must always be placed across the longest view of the room if they are to be effective. Suspended hollow beams containing tubes are yet another idea but here the bottom of the beam needs to be filled in with small-cell louvering which appears fairly bright, for one of the objects of the exercise is to attract the eye to the brightness so that it is not aware of what is going on above.

However, by far the cheapest method of apparently reducing the height of ceilings is to use a light colour on the walls up to the height you would like them to be with a darker colour above, and to hang lighting fittings with no upward light distribution level with the top of the lighter colour. The darker colour will need more thought here for in daylight it should not give a feeling of hovering doom. This is an effective treatment for narrow kitchens which have been sliced off larger rooms and you might well forget fluorescent fittings in favour of open cylinders with filament bulbs mounted on a suspended rod or batten as long as they don't get in the way of cupboard doors. Alternatively, you can have a ceiling-mounted wooden batten, the same colour as the ceiling, with fittings hanging from it. Grooved at the back to take the wiring, it can usually be easily screwed to joists and then there is no need for each fitting to have a ceiling plate or rose as the flex can come through neat holes in the batten.

Flexibility

In lighting jargon this refers to systems which allow you to change the position of fittings with the minimum of trouble, and not to the ability of things to bend. Various forms of 'light track' are available to which almost any sort of fitting can be 'clipped'. They can be attached horizontally and vertically to walls, flush mounted on, or suspended from ceilings and look thoroughly right in any 'architecty' interior. Most makes can be joined at right angles at any point and a track mounted on the ceiling about three feet away from the walls all round the room, or as a square in the centre, or in a single line with branches, could, with different types of fittings attached, take care of all the lighting. Different forms of track, usually referred to as 'trunking', are available for extending existing wiring. These are really intended for attaching to skirting boards and are a great help in providing extra socket outlets. One of the newest can be obtained in 10 feet lengths and is literally flexible to the extent that it can be taken round corners. It can be connected to 15 or 13 amp socket outlets and takes miniature plugs containing 3 amp fuses which makes it excellent for portable lamps, although it can also be used for any type of low-powered appliance. Both track and trunking can solve many wiring problems and it is worth balancing their cost against the more orthodox methods since one form or another may prove to be a real money-saver.

Fig 45 Wherever there is a 15 or 13 amp socket outlet, this system of surface wiring or 'trunking' can be used to provide extra socket outlets. When it is not the focal point as in the photograph it is remarkably inconspicuous.

Spotlights

If you really want to learn how versatile spotlighting is, go round any large modern store. Look at the windows, look at the displays and you will almost certainly find that everything that is particularly attracting your attention is spotlit, and not necessarily by built-in spotlights. Their uses are innumerable and while in most homes it is better to conceal the fitting, many of them are so well designed that for a

Fig 48 An attractive shadow effect is created on the ceiling between low-brightness inset panels by using an upward-pointing portable spotlight. *Author's home*

the ceiling will be every bit as effective as the light on the leaves. Spots need not be static. There is motorized equipment which moves a bulb through a predetermined arc and special projectors with revolving multi-coloured discs. Both are easily hidden and entertaining for special occasions, but for everyday use the unvarying movements become as boring as those in an artificial log or coal fire.

As host or hostess it may be well worth suffering a little glare for the sake of effect. For instance if you are pouring out drinks at a party a hidden spot could glamourize the bottles and glasses but be very glaring for you. This is the price you have to pay for effect, as anyone on the stage knows, but in such circumstances it would be wise to carry a hand mirror round the room beforehand and make sure that the glamorous lighting on things is not going to have disastrous effects on people – yourself in particular. Do watch this point always and not just in connexion with spots. Some people can sense when they're not looking their best and while it may be kind to give up your favourite armchair to an honoured guest it may not be equally kind to give him – or her – the full benefit of the high-powered lighting that, so rightly, you have arranged for reading.

Basements

A basement that is to be used during the day is the classic example of a room that requires a dual installation – one for the day, one for after dark – for nothing will make it feel more subterranean during the day than always having to have the after-dark lighting on. The aim should be to create a light room but not an obviously lit one, and this can hardly be better done than by using light colours on the walls and floodlighting them from sunny coloured fluorescent tubes behind pelmets, and recessing tubes in the ceiling above the window to reinforce what daylight there is. With a floor as light coloured as possible, the ceiling, which is best white even if it is a low one, will pick up sufficient light to give a sense of daylight. All the enlarging ideas like alcoves and niches help, too, and if they are lined in medium blue they will give almost a window effect. The deliberately false window can be remarkably successful and can be applied to any wall by building out a frame for concealing tubes at the top and sides to light a blown-up black and white or coloured photograph, but this should be seen through net curtains during the day and curtained in the ordinary way at night. It is, however, important to balance the lighting of the real and false windows to keep the natural effect, and arrangements for dimming the tubes would be useful here. If the room is too large for the side lighting to reach the centre, then inset black-hole fittings or low-brightness louver panels flush with the ceiling, with fluorescent tubes above, should be used. Unless the function of the room makes filament lighting thoroughly unsuitable it should be used after dark but, of course, as much of the daytime installation as looks happy and appropriate can still be kept on.

Period furnishing

Nearly all rooms are a jumble of mixed periods but most will have one particular overriding feeling which can serve as a guide. If you harbour them, throw overboard all sneers at 'imitation' lighting fittings, for if you have a room of any period before the introduction of electricity they will almost certainly be unavoidable. All the concealed lighting methods can be used. This is the technique employed in many museums and if concealment can be provided without harming the fabric of the building it is one which, from the antiquarian's point of view, could cause far less offence than compromising. Wall brackets and pendants will always be far more difficult to select than portable fittings since before the seventeenth century they did not exist in any form that can be satisfactorily imitated by electricity. Portables, on the other hand, are easy, and a simple guide to what will look right for the bases and standards is to consider chair and table legs and, during the time that four-posters were being used, bedposts. Pots and vases of a given period are also safe from the design point of view but the old cliché about 'good' designs, whatever their period, looking happy together still has a great deal of truth in it.

6 Exteriors

Exterior lighting can be strictly utilitarian, lighting house numbers, showing up steps, acting as a burglar deterrent and extending the day so that in spring and autumn you can carry on gardening long after it is dark. Or it can be magical. On the canvas of the night you can endlessly paint pictures with light that will astonish you with their unexpected beauty, for however used to exterior lighting you become it is always full of surprises. Change a colour here, alter the direction of the light somewhere else, and suddenly there is a new pattern of light and shade and colour that you would not have believed possible. But, as usual, first there are the practicalities to consider.

The major difference between an exterior lighting fitting and one intended for interior use is that the former must be weatherproof and it should continue to function and remain completely safe in the foulest conditions. In most cases this means that the bulb or tube is completely enclosed, and the enclosure sealed by a rubber ring or similar device. Although pressed glass (PAR) reflector bulbs are weatherproof in themselves, it is still essential that there is a sealing ring between the bulb and the holder to keep out moisture and the holder itself must be of a type designed for outdoor use. This cannot be stressed too strongly, for while it would be tragic to frighten you away from outside lighting it would be even more tragic if, by not being warned of the dangers, you received a bad shock or worse. Never run risks by using equipment that you are not sure about. Never take interior fittings out of doors, although occasionally some plastic shades can be used, and always ensure that everything you buy is designed for exterior use and installed by a qualified electrician who, while providing long cables so that you can place the lighting where you like, can quite easily ensure that everything *is* safe.

There is, for example, no reason why a ring circuit, equipped with fully waterproofed socket outlets, should not be installed in a garden. With the increasing use of electric lawn mowers, hedge trimmers and other garden appliances, a system of this kind has many uses apart from lighting and if the circuit is isolated from the mains by means of a transformer, voltages in the 230/240 range common in the UK can be brought down to far less lethal levels, and where lighting alone is being considered, those of almost complete safety. This will mean that lower voltage bulbs will have to be bought for exterior use but this is not at all a bad thing if it prevents unsuitable indoor equipment being used outside. Mind you, the necessity for the correct outdoor equipment, the correct earthing and the correct wiring still stands, however low the voltage.

Apart from having your own low voltage ring circuit installed, it is possible to buy complete kits consisting of one or two outdoor fittings operating on long cables from a transformer which can be connected to a mains voltage socket outlet. These are safe to use as long as the maximum number of fittings stated by the manufacturer is not exceeded and, most important, that the transformer is completely protected from the weather either indoors or in an outbuilding. Unfortunately this last point is not always made as clear as it might be.

Most people think of exterior lighting first of all as something for the front door, and it's as

Fig 49 Water is an enormous asset when lighting gardens. Here both white and coloured light are used.

good a starting place as any. The right fittings can be pleasant architectural ornaments by day and give a charming effect by night, but also make a real contribution to safety and security. Being able to light up unexpected late night callers by touching an interior switch at the front door can be a great comfort, and if the lighting is time-switch controlled it can act as a guard whether the house is empty or not.

The most popular fitting for front doors is the wall-bracket and a vast variety of different types and styles is available. Even when the interior is furnished and decorated in the modern idiom many people prefer a traditional fitting here, and it is surprising how well those of the carriage lamp type and genuine or reproduction gas street lamps can blend with modern architecture. Indeed this has led to many modern designs with the same sort of feeling but with the bulb only surrounded by a clear glass bowl. These look pleasant by day but, in visual terms, at night are no different from a bare bulb and a type with a diffusing bowl is usually preferable. A little light goes a long way out of doors and while a 40 watt bulb against a light wall may not appear uncomfortably bright, where such fittings with clear glass bowls are seen against dark backgrounds or used in the form of bollards for lighting paths – and how necessary it is that long paths should be lit – their glare may make their use almost as unpleasant and dangerous as having no lighting at all. Where the intention is to bring a certain amount of sparkle to the scene after dark, clear glass with a facetted surface is to be preferred, but with wattages above 40 a diffusing material around the light source is nearly always best. There are, however, many wall brackets of modern design with opal glass or plastic diffusers, some of which have flat surfaces to which the house number may be applied, and these are very suitable when there is no porch as such, or when a recessed porch area is of considerable size. In the first case the fitting can be mounted on the flat front wall of the house while in the second there is space to mount it inside the porch.

What is often unsatisfactory is a fitting on the wall outside a recessed porch so that quite a large part of the space within is left in shadow. The best solution here is to use self-effacing fittings or concealed lighting and the latter can be most effective in a glassed-in porch, although the late-caller problem should still be taken care of by a fitting over the outside door. If there is space for them, inset downlighters over plants at each side of the interior door are worth considering, while for general lighting the economy of fluorescent tubes is important for a light which may well stay on each night from dusk to bedtime. A 12 or 8 watt miniature tube has an almost negligible current consumption. Indeed some public authorities find that where the tube is used to light such things as 'No entry' signs it is simpler and cheaper to leave it on continuously rather than switching off during the day, taking into consideration, of course, the saving of the cost of a switch and the extremely long tube life that goes with uninterrupted burning.

Light around the back door is just as valuable. Being able to move in and out freely, to take rubbish to the dustbin, tools to the shed, or whatever it may be, is a blessing and gives that feeling of independence which is one of the functional aspects of exterior lighting. Simple sealed well-glass or bulkhead fittings are all that is needed, and one of the former, corner-mounted with a suitable bracket, can make light spread along two walls.

In nearly all exterior lighting watts are related to height. A large area can be lit by a small number of powerful light sources if they are mounted high enough, while lower mounting means more fittings with lower wattage bulbs or tubes. You could light a longish pathway, for example, by a single 100 watt or 150 watt PAR bulb under the eaves of a two storey house or by two or three fittings with 40 or 60 watt bulbs seven or eight feet above the ground or by eight or nine 15 watt bulbs or 8 watt tubes recessed into a wall at knee height.

When planning your wiring always include provision for internal as well as exterior lighting of outbuildings such as garages, greenhouses and sheds, and light them with thought. If there is a work bench in the garage a workmanlike 4′ fluorescent fitting should be mounted above and parallel to its front edge, and there should be another above the car engine as well as a socket outlet for a properly

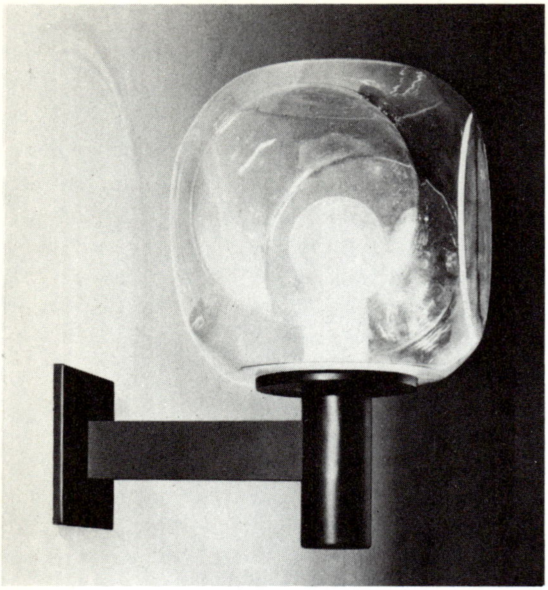

Fig 50 A well-designed wall bracket which can be obtained with clear, smoky or white glass. In the clear glass versions it is best not to exceed 40 watts unless they are being used on white or very light coloured walls.

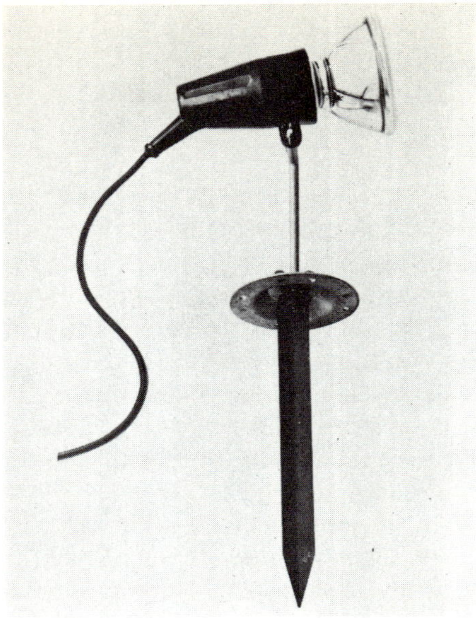

Fig 51 This fitting is one of the most useful for garden lighting. Not only can it be moved easily, but both the direction of the light and the height of the support can be adjusted.

earthed or low voltage hand lamp for inspection. Light outhouses so that you can find things and do whatever jobs you do there without getting in your own light, and give glass houses the kind of illumination that will enable you to see your seedlings and detect slugs and greenfly. All equipment here should be as waterproof and as soundly wired as if it were being used out of doors, but no real gardener will ever regret money spent on it for it means complete freedom from the restriction of darkness. So, of course, do fairly powerful bulbs with long cables on tall stakes which can be moved to any part of the garden. Use white cables if you can and keep them well off the ground. Two bulbs are always better than one if you are really to see what you are doing, for although a little light seems to go so far out of doors this is lighting for doing what may be a fairly detailed seeing task and 150 watt PAR bulbs should be the aim. This also applies to car cleaning and as well as lighting the approach to the garage and mounting a fitting above the door, a 4′ or 5′ fluorescent fitting each side of the car-cleaning position will make that job less of a Sunday morning chore. Such fittings should be mounted parallel with the sides of the car and high enough to light the roof. Economize by using only one fitting if you like but still keep it at one side and turn the car round – to mount it above the car would leave its sides in darkness.

When it comes to exterior lighting solely for effect there are three principal ways of going about it. One is to create a lighted setting for a lighted house – most rewarding if you're a show-off; another is to plan nocturnal pictures mainly to be seen from the interior or from a patio or loggia; and the third is to light the garden in such a way that it can be used and enjoyed by people in it.

For the first treatment floodlighting of the house is obvious. There are many excellent fittings for this, amongst the newest are 500 watt tungsten halogen floodlights which give more light and have longer lives than ordinary bulbs; and where it is possible to place one far enough away for its light to spread over the whole of the elevation to be lit, the rather high wattage is not an extravagance. But when considering floodlighting do not forget about windows. With rooms unlit and curtains open they look dead and dark, and while this can be corrected to a certain extent by closed curtains with light-coloured linings, it is worth considering how they can be made to form a lighted pattern of their own. One way is to install upward lighting inside at sill level (but this can be clumsy); another is to use thin curtain material and rely on pelmet lighting, making sure that the tubes are not visible from normal viewing positions; yet another treatment is to have exterior pelmets above each window. So much depends on the architecture, but if you have a house that will take them, the last suggestion can look delightful, particularly with window boxes. Indeed, a house lit solely in this way and with lighting at the front door can be remarkably effective with just enough lighting in the front garden for growing things to keep their identity without taking attention from the house. Never slosh light around outside just for the sake of it. For instance, it might be that a brightly floodlit light-coloured building would look all the better if there were no foreground lighting and trees seen only in silhouette. If you are lighting up just for Christmas you can hardly do better than festoon everything with strings of coloured lights. For a long time these were only made in the 40 or 60 watt bulb size, but now there are smaller ones and you may like them so much on the branches of a tree that you will keep them throughout the year and even the coloured glass balls that go so well with them.

When planning views to be seen from the interior any directional lighting must be pointed away from the house and all sources be hidden. Treat the garden as a stage. Pick out the good shapes; light upwards into trees; downwards on to flowers. Getting enough height is frequently a problem and placing fittings under the eaves may be the only solution where there are no tall trees. Lighting that is mainly to be seen through windows will have to hold its own with the room lighting, so fairly generous wattages will be needed; for this reason it may be better to concentrate on one or two brightly lit points of interest. Coloured light can be used, there are those who love it, but there is so much wonderful colour in a garden that it seems a pity to kill any of it, which any strongly coloured light will do. If you are using fluorescent tubes emphasize colour by all means with De luxe Natural or any of the De luxe colours but steer clear of Colour Matching, Northlight and high efficacy types.

Where the garden is to be enjoyed by people in it, it needs to be treated almost like a room – that is to say, no glare from any position and things and parts of it lit for use and effect. With really waterproof lamp holders some enclosed plastic shades with ordinary bulbs can be happily used as pendants in protected positions where they are not going to be swung violently by high winds. It is best, though, to keep to exterior fittings, of which there are many. Bracket fittings of all kinds are good where a fixing can be found and many are available in matching versions for post-top mounting. Of the free-standing types that can be stuck into the earth the 'mushroom' is probably best known. It is difficult, however, to use effectively for if its 'stem' is tall enough to light groups of flowers throughout the summer its bulb will nearly always be visible. This question of height is always cropping up and needs thought. Plants grow and it is sad to see a fitting gradually disappearing as the season progresses. Lighting ponds, streams, waterfalls and fountains is always rewarding and there is plenty of submersible equipment to be had. It should always be suitably screened by rocks or foliage but it will need frequent cleaning for algae do not mind the heat of the light sources and will very soon coat them completely. Fortunately fish are quite happy with them, too, although in a small pond it is wise to have an electric fountain to keep the water thoroughly aerated.

Any temptation to use portable equipment in swimming pools should be firmly resisted. If they are to have under water lighting it calls for a built-in professional installation, and if they are not suitable for this then fairly high-mounted surrounding lighting is all that should be provided. The vision of children playing in a plastic pool with floating lights may be charming but it is also quite terrifying and must never become a reality however low the voltage may be. For the lived-in garden, as well as permanent or semi-permanent lighting for barbecues, putting and tennis and croquet and all the other places which need the same lighting treatment throughout the year, there is still scope for varying effects and it is always fun to have in hand a few bulbs, both white and coloured, on stakes with long cables. It is with these that the extra magic can be created. However good a gardener you are, the unexpected is always happening, and to be able to light the early blossom if it is particularly lovely, to catch a laburnum during the all too short time it is at its best, a clump of lilies, a bank of iris, your roses – is to be able to make them visible and even more beautiful long after they would normally have faded into darkness.

7 Lighting definitions

The following are simple definitions of some of the technical sounding terms that have been used throughout this book, but most have been explained in the body of the text.

Amp (*short for ampere*) The rate of flow of electric current. Cables and wires must be thick enough to carry safely (i.e. without overheating) the flow of current normally drawn by the apparatus they serve. A fuse or circuit breaker is fitted into the circuit to interrupt it should the flow of current for any reason become excessive.

Batten fitting A slim white box-like container for all the necessary control gear for operating a fluorescent tube, carrying the lampholders which are usually hinged or sprung in some way so that the tube can be easily fitted. Most batten fittings are designed so that, if desired, reflectors or diffusers can be very simply attached.

Black-hole fitting A cylindrical fitting, horizontally grooved and painted black inside which can be recessed, semi-recessed or ceiling mounted. The design is such that unless one stands almost immediately beneath and looks directly upward it is nearly impossible to tell whether or not the bulb is switched on. The light in the room seems to come from nowhere.

Capacitor A component of the control gear for a fluorescent tube. It has no moving parts and can normally be expected to last as long as the fitting of which it is part. Its purpose is to assist economical operation.

Ceiling plate or rose The circular metal or plastic block within which the flex from the lighting fitting is connected to the fixed wiring installation. The word 'rose' comes from the plaster decoration around the suspension points of chandeliers which were often in the shape of radiating rose-like petals.

Choke The component of fluorescent tube control gear which limits the current to the correct value.

Circuit breaker Circuit breakers are now usually installed in the place of the old fashioned fuse boxes. They resemble switches which automatically turn themselves off if there is an electrical fault and should not be re-set until the fault has been traced and corrected. They are normally contained in neat boxes which for the average installation would measure about 12" x 6" x 3" deep.

Convection current The upward flow of air past any hot object. It carries dust with it and so is liable to cause a dirty patch above a wall bracket or on the ceiling above a pendant fitting. This will happen more quickly on cold surfaces where slight condensation helps the dust to stick and chimney-like shades are the worst culprits. The farther a fitting is away from a surface the less likely it is to occur and it can be cured to some extent by filling the bottom of the shade with a disc of clear plastic.

Diffuser Anything which surrounds the bulb or tube allowing light to pass through but softening it, such as white or frosted glass and plastic.

Earthing For safety from shock in the event of electrical or mechanical failure of electrical apparatus, the non-current-carrying metal parts which are liable to be handled should be 'earthed' *via* the earthing pin of a 3 pin plug and socket, or by connexion to metal which goes underground, such as a metal water pipe. If this is not done in surroundings which may be damp, or in situations where it may be possible to touch the apparatus with one hand and a water pipe, a bath, a radiator etc. with the other, then someone sometime may be severely hurt or worse. In some types of fluorescent tube circuits a good earth connexion is essential for quick and reliable starting of the tube, but whenever electrical safety depends on earthing it should be done professionally.

Electrical load The electrical load of, or the power taken by, a piece of apparatus. Wires and cables are of varying thickness (see flex) according to the assumed load that they are expected to carry and it is therefore unwise to increase the load to any great extent on a fixed lighting outlet without taking professional advice.

Flasher An automatic device which switches a circuit

on and off at frequent intervals, e.g. to obtain a scintillating effect with Christmas tree lights. Most simple flashers are of the thermal type which depend upon a bi-metal strip bending to break contact as soon as the current warms it up; when the contact is broken the current ceases, the strip cools and bends back again to remake the contact. More complicated flashers can be obtained for more sophisticated switching variations.

Flex Short for flexible sheathed current-carrying wiring normally used for connecting small-power apparatus such as lighting. There are many types and colours, some designed for heavier pendants incorporating an additional load-bearing wire. The thicker variations are usually referred to as cable. 'Cord' is also a word loosely used for both 'flex' and cables. Avoid knots or kinks in both in case the insulation is broken and never cut either without completely disconnecting from the mains. Kinky 'flex' can be straightened by immersion in hot water and quickly drawing it through a tightly held warm cloth. Make sure that the ends are dry before connecting.

Fused spur-box Looks just like a switched socket outlet without the holes for the plug pins. A spring release allows the fuse-carrying section to be easily removed without using a screw-driver.

Louver or louvre The former is the preferred spelling in the UK. A mesh or series of vertical fins below a light source to restrict the direction from which it can be seen and control distribution of light. Circular louvers are incorporated in many filament bulb fittings.

Lumen The measure of output of light from a source. Due to automatic adjustment of the sensitivity of the eye, a source having twice the lumen output of another will appear to be giving more light, but less than twice as much.

Lux (*illumination value*) The measure of the number of lumens arriving at a surface of given area - more or less equivalent to how thickly butter is spread on bread. Notice that this is quite different from the number of lumens *leaving* a source - many of them go anywhere but to the surface concerned. The metric unit of illumination is the lumen per square metre called lux. In the Imperial system, the unit is the lumen per sq ft or the synonymous American term 'foot-candle'. Very simply, 10 lux roughly equal one lumen per sq ft.

Opal glass and plastic White, or in a few cases coloured, translucent material used sometimes as a reflector but usually as a diffuser which, by increasing the area from which light is emitted, makes the source of light less bright and therefore more comfortable visually. Opal material is self-coloured; no external pigment, which could scratch off, is used to give it colour.

Prismatic material Prisms bend light. How much they bend it depends on the shape of the prisms and the angle at which light enters them. By careful optical design, it is possible to form one side of a sheet of clear glass or clear plastic into a continuous pattern of tiny prisms and to use this material in a lighting fitting to redirect the light in a predetermined way. Very often this principle is used to limit the amount of light going towards normal viewing positions and so to reduce glare. In the case of fluorescent fittings, the prisms are usually stretched into lines parallel with the tube axis, producing what is known as reeded glass (or plastic).

Resistor All materials used to conduct electricity will also to a greater or lesser degree resist its flow but will not stop it completely, e.g. the tungsten filament inside a bulb, the special wire used in the heating elements of fires, hotplates and so on.

The main domestic use of resistors is in radio and TV sets but very small ones are sometimes fitted inside the caps of miniature bulbs to effect the necessary control of current.

Rise-and-fall fitting A fitting incorporating in its suspension an arrangement of springs or weights which allow its height to be varied at a touch. Rise-and-fall suspensions are available on their own. The 'works' are sometimes contained in the ceiling rose, sometimes in a ball or egg-shaped enclosure on the suspension and in the counter-balanced type the weights and pulleys form an integral part of the fitting design.

Shade carrier The device which is held in position by the screw ring on the lampholder to support the shade. Carriers are available in several sizes and the correct device is necessary if the shade and base are to look right together.

Socket outlet The wall-mounted or inset component which receives the plug connecting to a piece of apparatus. Almost invariably all socket outlets in new installations are of the 13 amp variety with slots to receive the 3 pin flat-pin fused 13 amp plug; the plug holes in the socket are shuttered so that when the plug is withdrawn the current-carrying parts are not accessible. A 13 amp socket will deal with any load up to 3 kilowatts, such as a large electric fire, but the replaceable fuse in the plug must be of the correct rating for the piece of apparatus concerned. If the 'ring circuit' is properly and thoughtfully installed it should be possible subsequently to add further socket outlets with very little trouble. Each ring circuit is protected by a 30 amp fuse at the main switchboard. If this fuse blows, call expert opinion to find out why.

Starter switch A renewable and inexpensive component of many kinds of fluorescent fitting, which is essential to get the tube to light. If the tube flashes repeatedly when trying to start, or perhaps will not start properly at all, it is probably the starter switch which requires renewal.

Switches The ordinary (single-way) wall-mounted switch needs no comment, but three additional types are normally used:

1 Two-way switches. A pair of these, e.g. one in the hall and the other on the landing, is used to control, say, the hall light from either position, and another pair controls the landing light from either position, so that one need never go up or downstairs in darkness. Other possibilities for two-way switching are mentioned in the text.

2 Pull-cord switches. Like a ceiling rose with a hanging cord which is easy to renew if it gets soiled. Always pull sharply to avoid internal sparking which can damage the contacts. This is particularly important with fluorescent tubes.

3 Time switches. Time switches are essentially clocks incorporating electrical contacts, and are connected in a circuit to switch it on and off, or off and on at preselected times. Time switches are available which plug into a socket outlet; the appliance to be controlled is then plugged in to the time switch.

Transformer A small box-like device connected to the mains and used to produce a higher or lower voltage as may be appropriate for the apparatus concerned.

Translucent A translucent material scatters light which passes through it. When looking towards a tube through a translucent material such as figured or frosted or opal glass (or plastic) it will be difficult or impossible, depending on the degree of scatter, to distinguish the shape or precise position of the tube behind.

Transparent A transparent material such as clear window glass which transmits light without scatter.

Tungsten halogen A special variety of filament light source in which a halogen - iodine or bromine - is added to the gas filling. By careful bulb design, the halogen can be made to return to the filament the particles of tungsten which evaporate from it, thus prolonging its life and at the same time allowing it to be designed to give more light per watt.

Unit of electricity A power of 1000 watts is called a kilowatt (kW); one kilowatt used for one hour consumes one unit of electricity. If the wattage of the apparatus is multiplied by the hours of its use and divided by 1000, the result is the kilowatt hours - units of electricity - which have been consumed and it is on these that your electricity bill is based.

Ventilation Ventilation of lighting equipment is not generally of concern to householders since lighting fittings used in the normal way are designed to dissipate the heat generated by bulbs and tubes. It must be remembered, however, that in the case of built-in lighting, particularly where a fairly large wattage is concerned, ventilating holes to allow a through flow of air may be advisable to prevent an undue build-up of heat.

The control gear of fluorescent tubes generates some heat of its own which is normally dispersed by conduction or radiation from the fitting in which it is enclosed; but if, say, a batten fitting is cooped up in a small space from which heat cannot easily escape, or if the control gear is installed remotely in a similar situation, deliberate arrangements for ventilation should be made to avoid damage to components.

Voltage The electrical pressure which forces current through switched-on apparatus. With very few exceptions, domestic electrical supply in the UK is at a nominal 240 volts, but, unavoidably, the voltage at individual households may be a few per cent above or below this figure, depending on their distance from a sub-station and on whether or not their neighbourhood is at that time making a heavy demand on the supply.

Wattage The term of measurement for the electrical power taken by a piece of electrical apparatus. In the case of fluorescent tubes, the wattage of the whole circuit is the wattage of the tube plus that taken by the control gear - a 5' 65 watt tube circuit takes a total of about 80 watts.

Manufacturers

The items listed below are those that you may have difficulty in buying from non-specialized retailers. The manufacturers given will usually supply catalogues and information or the name of their nearest stockist. Those firms marked with an asterisk make most types of lighting fittings. Generally, however, it is better to look at the displays in good department stores and specialist lighting shops before approaching manufacturers.

1	Adjustable floor and table lamps	**19**	Luminous ceilings
2	Adjustable wall brackets	**20**	Metallized louvering
3	Art nouveau shades	**21**	Modern 'crystal' fittings
4	Black-hole fittings	**22**	Motorized spots
5	Bulbs and tubes	**23**	Multi-way switching
6	Christmas tree bulbs	**24**	Night lights
7	Circular fluorescent fittings	**25**	Picture spots
8	Clip-on louvers	**26**	Rhythmic lighting
9	Dimmers	**27**	Rise-and-fall fittings
10	Electrical accessories	**28**	Shade carriers
11	Exterior fittings	**29**	Shaver-light units
12	Filament table lamps	**30**	Special candle bulbs
13	Flashers	**31**	Spotlights
14	Fluorescent table lamps	**32**	Surface trunking
15	Heat and light fittings	**33**	Time switches
16	Lightmeters	**34**	Three-light bulbs
17	Light-track	**35**	Traditional fittings
18	Low voltage miniature lamps	**36**	Translucent material

Abbey Electrical Systems Ltd, 95 Victoria Street, St Albans, Herts **32**
Adel Roostein (Display Sales) Ltd, 17 Soho Square, London WI **3**
I. M. Allen & Co Ltd, 1 Links Green Way, Cobham, Surrey **35**
* Allom Heffer & Co Ltd, 17 Montpelier Street, London SW7 **4, 7, 30**
Arrow Plastics Ltd, Arrow Works, Hampden Road, Kingston-on-Thames, Surrey **36**
Avo Ltd, 92-96 Vauxhall Bridge Road, London SW1 **16**

Baker & Finnemore Ltd, 197-207 Newhall Street, Birmingham **28**
Barries Electrical Agencies Ltd, Canning Street, Brighton, Sussex **10**
* BBI Ltd, 58-60 Berners Street, London W1 **9, 11, 17, 19, 27**
* Benjamin Electric Ltd, Station House, Darkes Lane, Potters Bar, Herts **11**
BICC Ltd, 21 Bloomsbury Street, London WC1 **10**
British Electric Lamps Ltd, Elm Grove, London SW19 **30**
Bright Light Installations Ltd, 6 Avenue Road, London NW10 **7**

Chilton Electrical Products Ltd, Hungerford, Berks **29**
* C. M. Churchouse Ltd, Lichfield Road, Brownshills, Walsall, Staffs **11**
* Clarke (Electrical) Ltd, 253 Kilburn Lane, London W10 **21**
E. C. Colton (Connectors) Ltd, Aetna Works, Moor Lane, Craham, Upminster, Essex **10**
Combex Ltd, 117-123 Great Portland Street, London W1 **24**
* Concord Lighting International Ltd, Rotaflex House, City Road, London EC1 **1, 2, 4, 7, 11, 17, 18, 24, 25**
* Conelight Ltd, Longmead, Shaftesbury, Dorset **2, 7, 31**
Contromatic Ltd, Moor Street, Brierley Hill, Staffs **33**
Cooper-Wallace Ltd, 63 East Barnet Road, New Barnet, Herts **13**
* Courtney, Pope (Electrical) Ltd, Amhurst Park Works, London N15 **4, 7, 11, 18, 19, 20, 31, 32**
Crabtree & Co Ltd, Walsall, Staffs **10, 33**
Cressall Ltd, Cheston Road, Birmingham 7 **9**
Crestworth Ltd, 59 Kennington Road, London SE1 **26**
* Crompton Parkinson Ltd, 39-41 Bridge Street, Northampton **11**
Crysaline Ltd, Waterside, Chesham, Bucks **1**
Cryselco Ltd, Kempston Works, Bedford **7**

Dawe Lightomation Ltd, South Hill Lodge, South Hill Avenue, Harrow, Middlesex **26**
* Decorative Lighting Manufacturers and Distributors Association, Members of, Sardinia House, Sardinia Street, London WC2 **12**
Delmatic Ltd, 52 Upper Berkeley Street, London W1 **23**
Dernier & Hamlyn Ltd, 23 Newman Street, London W1 **35**
Designplan Lighting Ltd, 18 Grove Road, Sutton, Surrey **11**
Dorman Smith Ltd, Preston, Lancs **10**

Elco Plastics Ltd, Desborough Park Works, High Wycombe, Bucks **20**
* H. H. Electrical Ltd, 253 Kilburn Lane, London W10 **29**
Electrosonic Ltd, 173 Greenwich High Road, London SE10 **9**
Electronics Ltd, Broadgate Lane, Horsforth, Leeds **9**
Elkay Electrical Manufacturing Co Ltd, 51 Weston Street, London SE1 **10**
Elsworthy Electronics Ltd, 27-31 Broadley Terrace, London NW1 **11**
Emess Manufacturing Co, 30-32 St James Street, London E17 **11**
Enicorn Industrial Devices Ltd, 9a Annette Road, London N7 **22**
Ever-lite Electric Co, 93-97 Regent Street, London W1 **26, 30**

Falcon Forge Ltd, 33 Avery Hill, New Eltham, London SE19 **11**
* Falks Ltd, 91 Farringdon Road, London EC1 **2, 4, 8, 9, 11, 27, 31, 33, 35**
C & R Flashers Ltd, 42 Thames Street, Hampton, Middlesex **13**
Fraser Field (Sales) Ltd, Royston, Herts **11**

* Grubert of Copenhagen Ltd, 293-299 Kentish Town Road, London NW5 **2**
Guy Puckle & Co Ltd, Warnford Court, Throgmorten Street, London EC2 **25**

Heaton & Boyes Ltd, 11-13 Millbank Street, Manchester 1 **22**
* Holophane Ltd, PO Box 36, Bond Avenue, Bletchley, Bucks **36**

Impex Ltd, 26-30 Holborn Viaduct, London EC1 **35**
Inca Construction Co Ltd, Wycombe Road, Stokenchurch, High Wycombe, Bucks **36**
D. R. Illuminations Ltd, Harper Road, Wythenshawe, Manchester 22 **11**
Isora Integrated Ceilings Ltd, Buckingham Avenue West, Slough Bucks **19**

Joseph Lucas Ltd, Great King Street, Birmingham 19 **15**

Lennig Chemicals Ltd, Lennig House, 2 Masons Avenue, Croydon, Surrey **36**
* Lighting Industry Federation, Members of, 25 Bedford Square, London WC1 **5**
* Linolite Ltd, 142 Liverpool Road, London N1 **29**
* Lumitron Ltd, Hythe Road, London NW10 **18, 22**